Site and Sound

Site and Sound

*The Architecture and Acoustics
of New Opera Houses and Concert Halls*

Victoria Newhouse

The Monacelli Press

For C.

Copyright © 2012 by The Monacelli Press LLC

All rights reserved. Published in the United
States by The Monacelli Press, New York

Library of Congress
Cataloging-in-Publication Data
Newhouse, Victoria.
Site and sound : the architecture and
acoustics of new opera houses and concert
halls / Victoria Newhouse. — 1st ed.
p. cm.
Includes bibliographical references
and index.
ISBN 978-1-58093-281-3
1. Theater architecture. 2. Music-halls.
3. Architectural acoustics. I. Title.
II. Title: Architecture and acoustics of
new opera houses and concert halls.
NA6821.N45 2011
725'.81—dc23   2011043178

Printed in China

www.monacellipress.com

10 9 8 7 6 5 4 3 2 1
First edition

Designed by Pentagram

Acknowledgments

Site and Sound owes its existence to the opera director Gerard Mortier, who suggested that I turn my attention from art museums—about which I have written extensively—to opera houses and concert halls. Not only did the idea present a fascinating new subject, but it introduced me to contemporary classical music and to the many stunning new buildings for music being built around the globe.

I have relied on professionals for assessments of acoustic quality. No single person has been more helpful in this respect than J. Christopher Jaffe of JaffeHolden Acoustics (1967–2009) and Acentech (since 2009), who has been tireless in explaining acoustic theory and answering endless questions. Jaffe also introduced me to his colleague at Acentech, Carl J. Rosenberg, who accompanied me on my survey of theaters in China and astutely explained why performances in different spaces sounded the way they did.

Many other acousticians have also been of great assistance, including Robert Essert of Sound Space Design, England; Peter Fearnside of Marshall Day Acoustics, Australia; Eckard Mommertz and Jörg Kümmel of Müller-BBM, Germany; Professor Renz L. C. J. Van Luxemburg, architectural acoustics, Eindhoven University of Technology, Netherlands; Raj Patel and the late David Taylor of Arup, New York; Yasuhisa Toyota of Nagata Acoustics, Los Angeles; Toshiko Fukuchi of Nagata Acoustics, Tokyo; and Fred R. Vogler of Sonitus, Los Angeles.

The trip to China would have been meaningless without the expert guidance of Joanna C. Lee, Museworks Limited, Hong Kong. She arranged meetings with key personalities in the nation's music world and, together with her husband, Ken Smith (music critic for the *Financial Times*), read and critiqued the manuscript.

Other readers of parts or all of the text who have made invaluable suggestions are Larry Busbea, David De Long, Nick Frisch, Neal Goren, Alex Gorlin, Jaffe, Sissy List, Gerard Mortier, Sato Moughalian, Rosenberg, Mildred Schmertz, Suzanne Stephens, and my husband, Si.

Among the many musicians who shared with me their reactions to sites they had experienced as performers or listeners, Raphael Mostel played an essential role. I first encountered this composer and critic when I audited his Columbia University seminar "Architectonics of Music," taught in tandem with Steven Holl's advanced studio course in architecture. This book could not have been written without Mostel's thought-provoking guidance.

As for my last two books, Robert Gottlieb helped me shape a formless mass of material into a coherent whole. Andrea Monfried has been, as usual, a brilliant editor, and Terry Bissell an equally efficient copy editor. Rose Merola has been indefatigable in locating and obtaining images, and the quality of these images was ensured thanks to the technical assistance of Jack Lindholm. I am indebted to Elizabeth White for her expert production of the book and to Yve Ludwig of Pentagram for its design.

Introduction

SINCE THEIR ORIGINS IN ANTIQUITY AS PART OF RELIGIOUS RITUALS, THEATERS have played a pivotal role in society. Shortly after their appearance as places of entertainment in the early Baroque era, theaters for music—opera houses—became focal points of social and political life.[1] Following the first opera houses by approximately a half-century, concert halls were also favorite places of social intercourse, as both are today.

Operas and concerts are relative newcomers to the age-old history of theater. *La Dafne* by Jacopo Peri, generally considered to be the first true opera, premiered in Florence in 1598. It was probably performed at the home of Jacopo Corsi, a prominent patron of music who collaborated on the work. About a hundred years passed before instrumental concert music began to develop; it too was presented in private venues.

When architects started to design public spaces for these new genres, they established interior forms that have for the most part remained in place for centuries: the horseshoe plan for opera, which has roots in ancient Greece and was developed in mid-seventeenth-century Venice, and the shoebox shape for concerts, which evolved toward the end of the seventeenth century primarily from ballrooms and riding schools in Germany and the Austro-Hungarian Empire. It is the specter of flawed acoustics that has frozen these historic forms in time. While the exteriors of buildings for music became increasingly spectacular, their interiors remained, until the 1960s and even to the present, largely unchanged. (The one exception is the interior of Richard Wagner's revolutionary Festspielhaus of 1876 in Bayreuth.)

The billowing sail-like forms of Jørn Utzon's Sydney Opera House (1973) and the dramatically cantilevered roof of Jean Nouvel's Culture and Convention Centre, or KKL, in Lucerne (2000) both contain traditional theater layouts. More recently, the striking exteriors of the Oslo Opera House by Snøhetta (2008) belie a conventional interior. Only in recent decades have a small number of architects begun to deviate from these tried and true formulas. There may be no other building type with so great a disjunction between outside and inside.

A chief concern for historians of architecture, like myself, is the evolution of form. The near absence of such development for the performance spaces in opera

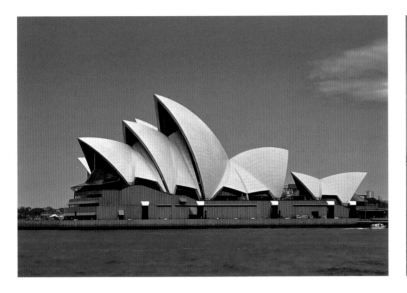

houses and concert halls is intriguing, and the selection included here was driven by my search for innovation within the limitations imposed by acoustical considerations. Frank Gehry's description of acoustics as "mystical magic straight out of *Alice in Wonderland*"[2] was an additional provocation. Coming from the architect of several successful indoor and outdoor performance areas, the statement raised basic questions: is acoustics an ephemeral art or a predictable science? And tangentially, does music serve the space or is it the other way around?[3]

Whether music is heard in a home, church, cathedral, palace, concert hall, opera house, or even in the open, no single factor has a greater influence on how it is performed and perceived than acoustics. And yet it is not uncommon for two experts to propose radically different solutions to a problem: there is no infallible prescription for acoustical excellence, though some experts deny this.[4] Nor can there be a definitive judgment about the acoustical quality of a given space: skillful musicians can temper problematic conditions, and every listener has a personal preference.

The natural (unamplified) acoustics of a particular space refers to how sound waves operate there: the time a sound continues, or resounds, after it is made, and its characteristics. Acoustics are affected primarily by the size, shape, interior design, and materials of an environment. Sound waves first reach the listener directly and then by reflection from surrounding surfaces. Human bodies absorb sound waves, as do other soft surfaces. Even clothes can influence reflections by muffling sound waves; many auditoriums require coats to be checked for just this reason. The sound during a rehearsal, when a hall is for the most part empty, can be quite different from the sound during a performance, when a full audience is in place. Furthermore,

Sydney Opera House, Sydney, Australia, 1973.

Culture and Convention Centre (KKL), Lucerne, Switzerland, 2000.

acoustics are important not only to the audience: the musicians must also be able to hear one another across the performance area.

Until the end of the eighteenth century, music was typically composed locally, for an existing interior: composers and the architecture for which they devised a piece were of the same era. In the nineteenth century, bigger orchestras, grander music, and programs that began to include compositions from a variety of historic periods did well acoustically in theaters that had grown much larger to accommodate both these musical changes and audiences that were greatly expanded by the rising middle and merchant classes. Acoustical challenges widened in the twentieth century when theaters began to serve a number of musical genres.[5]

A relatively small change within the basilica of San Marco in Venice provides an illuminating early example of how architecture and acoustics can interact. A middle Byzantine masterpiece inspired by Justinian's Church of the Holy Apostles in Constantinople, the building has fine acoustical properties—to which its five domes contribute—that are uniquely suited to the endless series of religious and civic celebrations, all featuring music, for which Venice is famous.

The polychoral style (spatially separated groups of singers) originated at San Marco under Adrian Willaert, *maestro di cappella* of the basilica from 1527 to 1562. In the middle of the sixteenth century, the architect-sculptor Jacopo Sansovino inserted two small balconies for vocalists at either side of the cathedral's existing chancel;[6] the alteration was so effective acoustically that it inspired Willaert's followers, Andrea Gabrieli and his nephew Giovanni, to create monumental polychoral works there, ever since associated with the glories of this city.[7]

Remarkably, in the twenty-first century, the models of acoustical perfection remain concert halls that were constructed more than a hundred years ago. The Musikverein in Vienna (1870) and the Concertgebouw in Amsterdam (1886), when occupied, have measured reverberation times of approximately two seconds in

Basilica of San Marco, Venice, Italy. Views of chancel with Jacopo Sansovino balconies, mid-1500s.

mid-frequencies,[8] an ideal that can be compromised, for cognoscenti, by a fraction of a second. Reverberation time is, of course, not the sole consideration: clarity, warmth, and balance of sound also play a part in the acoustic quality of a given place. And evaluation is meaningful only in relation to a specific type of program.

The Musikverein and the Concertgebouw both enjoy extraordinary quality of sound, but they are different from one another, and neither is ideal for everything. The Vienna hall's narrow shape provides strong rear and lateral reflections, creating an intimate feeling; in the wider Amsterdam hall, the reflections are not as forceful, offering a more spacious and less enveloping effect. Experts judge the long reverberation time of both auditoriums ideal for Romantic works; the Musikverein does well with Brahms and some Mahler, the Concertgebouw with big Mahler pieces.[9] The two halls are less well-suited to most Baroque, classical, and a great deal of modern music. Indeed, I noticed an uncharacteristic lack of clarity at the performance of Esa-Pekka Salonen's own work, which he conducted in the Grosser Musikvereinssaal.[10] In addition to the acknowledged problems presented by some of today's music, it is difficult to imagine in either space the playing of certain contemporary multimedia compositions, which might include newly invented and amplified electronic instruments, not to mention video projections and dramatic lighting effects.

Acoustical excellence grew even more elusive in the 1960s, when it became apparent that individual theaters dedicated to a single art form were far too costly for most communities to undertake. In order to survive financially, opera houses and concert halls must be available for rent when their home companies are not in residence. So in addition to the natural acoustics needed for classical music from various eras, these theaters also have to provide the electronic amplification systems that are necessary for a variety of other genres, such as popular music, dance, and film, plus the electronics utilized in contemporary compositions. And there is a third option: what I refer to throughout this book as electronic architecture, systems that process and transform sounds in a number of ways, usually to mimic the acoustic qualities of different spaces. A frequent compromise—and the most different from theaters of the past—is a multiuse theater intended to serve an array of performing arts.[11]

Despite the present-day rarity of a one-to-one relationship between architecture and music composed specifically for it, many of today's music venues offer visual and aural conditions that inspire composers and please audiences. In order to combine, and give equal importance to, judgments about the architecture and acoustics of opera houses and concert halls of the past decade, I have relied on professional acousticians and music critics.

The nature of acoustics and its relationship to architecture are not the only questions posed in *Site and Sound*. There is also the puzzling paradox of a worldwide boom in the construction of music facilities at a time of what appears to be declining attendance and aging audiences, together with the preference of many young people for less formal environments. Nevertheless, these costly, high-profile buildings are replacing museums as linchpins of urban expansion and tools of global politics and cultural economics. Like many art museums, the most ambitious new opera houses and concert halls are the twenty-first-century equivalent of the medieval cathedral: they have been shorn of their reverential aura, but their magnificence remains a symbol of wealth and power. Swings in users' attitudes—between exclusivity and inclusiveness—have influenced the architecture of opera houses and concert halls for centuries. The present crop of increasingly transparent, more inviting theaters reflects the democratization of art that has occurred since 1950. They replace earlier generations of intimidating, templelike forms. Whether these icons will share the same kind of success that has been enjoyed since the late 1970s by museums remains to be seen.

Chapter 1, "The Past: A Historical Overview," looks at the origins of theater architecture, in ancient Greece and Rome, and its development in modern times, beginning with the Renaissance and continuing to the present. Carlo Fontana's balconied horseshoe at the Teatro SS. Giovanni e Paolo in Venice (1654) was challenged by Wagner's Bayreuth Festspielhaus (1876) for opera, just as Hans Scharoun's Berlin Philharmonie (1963) challenged the shoebox for concerts. Walter Gropius's unbuilt Total Theater (1927) introduced to the twentieth century an alternative to both—a reconfigurable, modular auditorium. Another nontraditional modern theater was the temporary structure for which specific music was composed, exemplified notably by Le Corbusier and Iannis Xenakis's Philips Pavilion for the 1958 Brussels World's Fair.

Chapter 2, "Lincoln Center: 'From Behind Walls to the Street,'" provides a link between the recent past and the present by comparing current renovations at Manhattan's Lincoln Center for the Performing Arts with what was built originally. This chapter encapsulates the shifts in goals, tastes, urban planning intentions, and attitudes toward the arts that have taken place in the United States since the early 1960s. Also examined are the ways in which acoustical considerations, and working methods generally, differ today from those of the mid-twentieth century.

Chapter 3, "The Present: 'If You Can Step on Something You Feel You Own It,'" offers an in-depth description of outstanding new opera houses and concert halls in North America and Europe. From Snøhetta's startling, hill-like Oslo Opera House to the ephemeral textile ribbon of Zaha Hadid's small, temporary JS Bach Chamber Music Hall (2009), these spaces offer an intimacy that is as important as the quality of their acoustics.

Chapter 4, "China: 'Building Big,'" explores the multihall Grand Theaters that are rising in the People's Republic at an astonishing rate. Part of the Chinese government's massive construction program to boost culture, these halls are symbols of regional political power much as nineteenth-century Western cultural institutions were built as emblems of a newfound nationalism. The usual rules and criteria do not apply in this ancient land, currently reinventing itself, where the possibilities seem endless.

Chapter 5, "The Future: Near and Far," reveals a new generation of opera houses and concert halls worldwide, whether just completed, currently under construction, planned, or merely hoped for. Europe and some East Asian countries have left the United States far behind in terms of experimentation, often inspired by the concept of a modular theater.

Offsetting the negative reports of dwindling audiences for traditional performances and reduced government support worldwide are the considerable successes of privately subsidized, tuition-free educational facilities and the proliferation of alternative performance spaces. Furthermore, a global outpouring of significant new operas and instrumental works speaks of lively, wide-ranging activities waiting to be captured by new sites for sound.

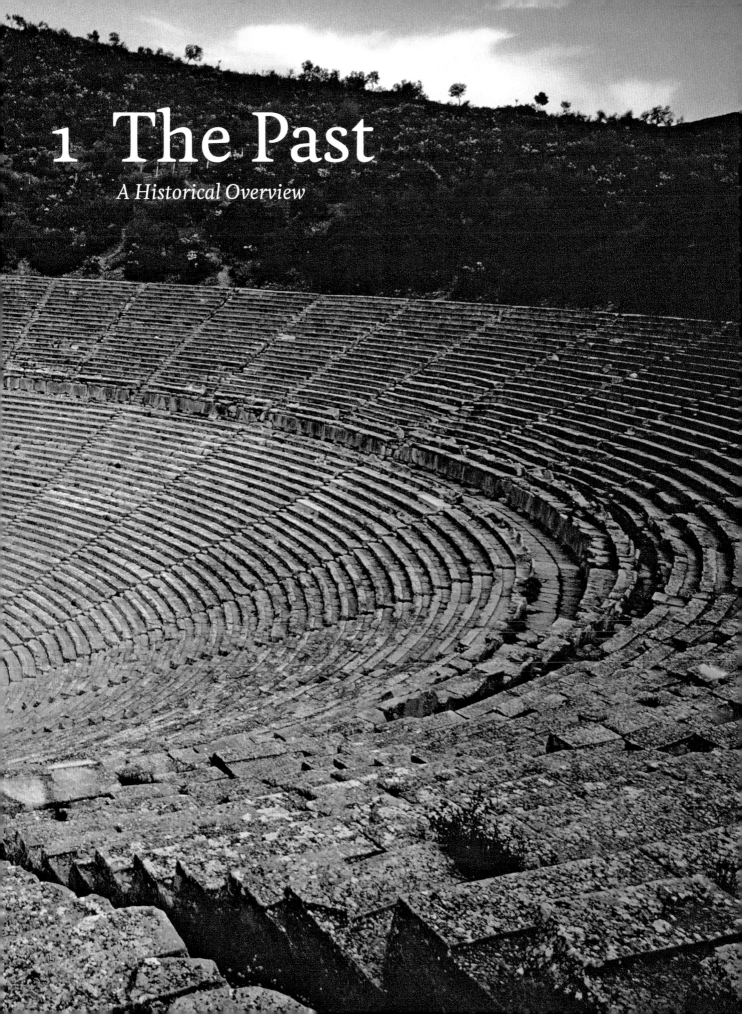

1 The Past

A Historical Overview

T

HE BIRTH OF OPERA OCCURRED TOWARD THE END OF THE
sixteenth century.[1] At the time Florentine intellectuals, convinced that music could
intensify the emotions of poetry, attempted to reinvent Greek drama, believed to
have been orated in a stylized manner to the accompaniment of chanted or sung
choruses and instrumental music. Less than half a century later, opera was being
presented in specially designed theaters related to those of antiquity. In a parallel,
though subsequent, development, opera's orchestral episodes led to concert works
that were eventually performed in their own uniquely derived spaces.

Western theater architecture originated in Greece during the sixth century
BCE and was developed subsequently by the Romans. Renaissance architects modeled their performance spaces on these antecedents, establishing the roots of

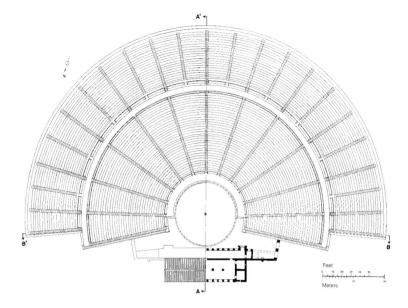

contemporary theater and opera house design. Greek drama seems to have evolved from the collective religious rituals regularly performed in honor of the god Dionysus, such as the Athens festival, City Dionysia. In Athens, where tragedy and comedy were born, and eventually in other cities as well, theater was intrinsic to the most important religious festivals, and therefore a major form of social intercourse. Popular dramatic productions were among the celebrations that conveyed the history and character of Athens, with theaters (along with the agora and the assembly) gradually replacing temples as the main architectural expression of Greek urban life. (In Rome's Imperial epoch, public baths also became one of these representative building types.)[2]

The form of the theaters was inspired by the simple premise that if the source of a sound is reasonably close and can be seen, then that sound should normally also be audible. Raked seating that curved around a flat, circular performance space (called *orchestra* by the Greeks) fulfilled the requirement.[3] The arrangement brought audiences in the thousands as close as possible to the action, so that they embraced the acting. The temporary wooden bleachers of early theaters were eventually replaced by permanent stone seats and risers arrayed in a semicircle, as exemplified by the Theatre of Dionysus (c. 325 BCE), dedicated by the orator Lycurgus on the south slope of the Athenian Acropolis and significantly modified since then.

The well-preserved outdoor theater in the northeastern Peloponnesus at Epidaurus, noted for its legendary acoustics, is critical to an understanding of the original appearance of similar Hellenistic sites. Still used for theatrical presentations, it was built at the beginning of the third century BCE. Symmetrically curved, and seating about twelve thousand, it is set into a hillside and divided by stairways. Performances initially took place in the orchestra, the front of which was marked by a circular stone, possibly the former base of an altar. Behind this area was the *skene*, at first a tent, then a wooden hut where the actors changed their masks and costumes; it, together with the wooden stage, also came in time to be made of stone.

Instead of the hillside sites into which the Greeks fitted stadium-type seating, the Romans preferred freestanding structures on level ground, still open to the air, but eventually with their curved peripheries closed to the exterior. Roman theaters shed the religious associations of their Greek counterparts and became part of civic and entertainment complexes that often included a forum and a stadium. The circular orchestra became a semicircle, thereby reducing the extent to which the audience surrounded the action. By the late third or early second century BCE, performances had moved to an elevated stage, which improved sight lines and sound, and also separated the circulation of actors and audience. But at the same time, the change distanced the action still further from the spectators, diminishing what had been the nearly theater in the round of earlier sites. The skene (called *scaenae frons* by the Romans), already expanded in Hellenistic times, was technically updated and extravagantly embellished with statuary and other adornments, usually representing a palace facade. Stages were enriched first with silver, then gold, and even ivory, and shade was provided by fabulously decorated awnings (*velas*).[4]

Only in 55 BCE did the Romans construct their first theater entirely in stone. The Theatre of Pompey was erected just outside the formal boundaries of Rome at that time, next to the Campo de' Fiori, by the triumphant general Pompey in the second year of his consulship. The building boasted boxes for the privileged, and the encompassing "leisure-complex" of gardens included a portico under which paintings and sculpture were exhibited. The huge—reputedly forty-thousand-seat—auditorium was the first example of Imperial architecture. Heralding Roman identity and ideology, it became a prototype for theaters throughout the Empire.[5]

In addition to the tiered, fan-shaped seating arrangements that have survived in modern theaters, Roman structures have passed down to the present another characteristic: the ability to reconfigure themselves and serve a variety of functions. In this they anticipated twentieth-century multiusage. Two temporary theaters built in wood in 52 BCE by C. Scribonius Curio turned on pivots. When back to back, they were used for performances; when face to face, they formed an amphitheater that held the larger audience attracted by gladiatorial combats.[6]

Theatre of Dionysus, Athens, Greece, fourth–first century BCE. Computer model.

Roman theater, first century BCE. Computer model.

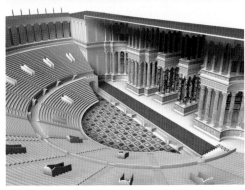

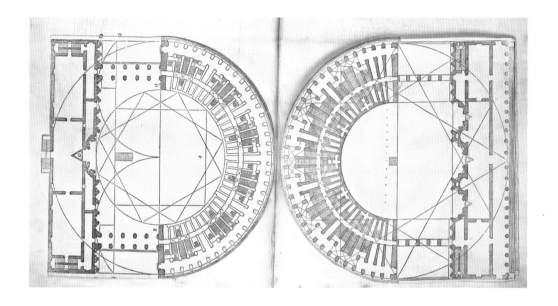

Theater of C. Scribonius
Curio, Rome, Italy, 52 BCE.
From Barbaro's *Vitruvius*.

Music was intrinsic to daily life in Greco-Roman antiquity, and the playing and enjoyment of music on its own was one of several activities that took place in a building called the *odeion*—or in Latin, *odeum*—a typology that developed alongside the *theatron*, which housed productions of tragedy and comedy.[7] From the middle of the fifth century BCE on, these roofed odeions rose near the more prevalent outdoor venues: that of Pericles, for instance, was erected next to the as-yet-undeveloped Theatre of Dionysus in Athens. Acoustics were generally superior in the relatively small indoor theaters to those outdoors: Plutarch is among the ancient literary sources that mention musical contests in the Odeion of Pericles (c. 446–42 BCE) at the time of the Panathenaic festival.[8]

An odeion is characterized by rectilinear outer walls, and the one built by Pericles, which likely seated up to four thousand, is thought to have been square. Lighting probably came from a central clerestory lantern augmented by windows in the side walls.[9] Built mostly of timber, the interior was filled with a grid of structural columns, which in addition to the predominantly flat floor and absence of a raised stage would have impaired sight lines—and possibly acoustics as well, because of the multiple reflections they would have caused.

Remains of the rectangular Odeum of Agrippa (16–14 BCE) in the Athenian agora demonstrate the degree to which the Romans refined this kind of assembly hall, as they had Greek theaters. The structure, without exterior decorative sculpture, was built of limestone and the same finely dressed marble as the Acropolis buildings.[10] The limited width of the room—enforced by the ninety-foot maximum length of the

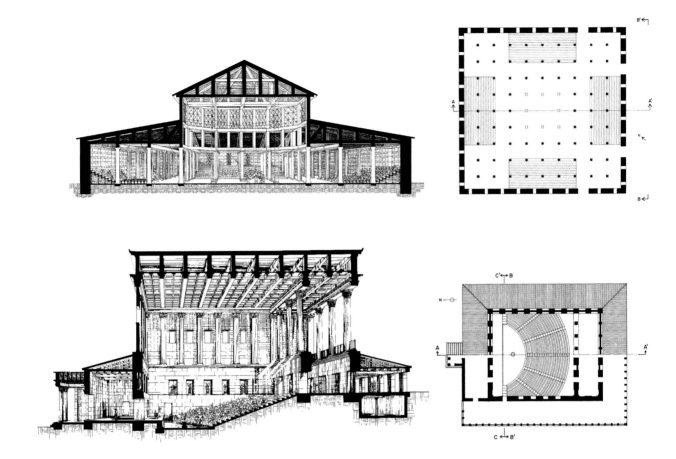

Odeion of Pericles, Athens, Greece, mid-fifth century BCE. Imagined section and plan.

Odeum of Agrippa, Athens, Greece, 16–14 BCE. Imagined section and plan.

wood roof trusses then available—eliminated the need for column supports, thereby improving sight lines and acoustics. About a thousand seats rose in curved tiers from a flat orchestra. Interior walls were articulated with marble inlay, the upper portions strengthened with pilasters that may have alternated with windows. Openings on the side and rear walls of these theaters provided light and air; they also served an acoustical purpose by reducing reverberation time, thereby improving speech intelligibility and the clarity of small musical ensembles.[11] Details, possibly including a deep coffered-wood ceiling and statuary, were extremely elegant. Open spaces between the Corinthian pilasters on the south wall provided the first known instance of direct front lighting for a stage, which in this case was raised. The Roman odeum may have been an elitist alternative to increasingly unwieldy, though popular, outdoor theaters.[12] The containment of its tiered, fan-shaped seating within a roofed, clear-span, rectangular container anticipated the modern theater.

The Opera House

The Renaissance architect Andrea Palladio's interest in the past brought the Roman theater into the modern era. His Teatro Olimpico in Vicenza was one of the first enduring roofed theaters after a hiatus in theater construction of nearly a thousand years caused by Church opposition to this form of entertainment. It remains intact and in use, primarily for spoken drama, although in at least one instance, accompanied by music.[13] Like many of his scholarly contemporaries, Palladio immersed himself in the lessons of antiquity. Not only did he visit Roman ruins, he studied Vitruvius's *De architectura*, the sole surviving architectural treatise from the ancient world. Palladio's design for the famous, approximately four-hundred-seat theater in Vicenza (completed in part by his son Silla di Pietro in 1583 after the architect's death, and finalized in 1585 by Palladio's pupil Vincenzo Scamozzi) is based on Vitruvius's description of a semicircular Roman theater. The scaenae frons (also by Scamozzi) is typical of such a building; the seating arrangement, size of the stage, and windows are more akin to an odeum.[14] Like other theaters at this time, the Teatro Olimpico was constructed entirely of wood (a preferred material for auditorium interiors to this day) within an existing brick structure, the Castello del Territorio, originally a fortress.

Palladio's revival of a Roman forerunner influenced subsequent designs by others, but it had few true imitators. The Teatro Olimpico's wide, semi-elliptical seating (made necessary by site restrictions) and open-platform stage were rendered obsolete by the inauguration in 1628 of Giovanni Battista Aleotti's three-thousand-seat Teatro Farnese for the Duke of Parma. It too was inserted into an existing building, but in this case the duke's Palazzo della Pilotta, a private rather than a public building, was used. Completed in 1618, the wood and painted stucco theater occupied a long, narrow former armory; seating arranged in a deep U marked the beginning of the longstanding horseshoe shape for opera house interiors. Raked rows faced a proscenium, picture-frame stage (in vogue since the early sixteenth century) with movable mechanized scenery, an arrangement that was generally preferred to the multiple prosceniums of the Palladian scaenae frons. Action was not

Teatro Olimpico, Vicenza, Italy, 1583. Interior view; plan.

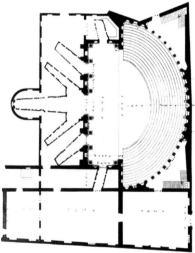

confined to the stage, however, but could spill out into the middle of the U, and the ability of Aleotti's theater to serve cavalry jousts, aquatics (by flooding the central arena), and balls, in addition to opera, was greatly prized. Within a few years, every self-respecting Italian court had its own permanent, private theater.[15]

In contrast to the democratic nature of theater in Greece, Renaissance opera was limited to a privileged few, lending it an aura of exclusivity that has remained associated with the genre to this day. Operas presented in the private theater and gardens of the Farnese palace in Parma, like those of the Medici in Florence, were elaborate spectacles enjoyed only by the elite. Well into the eighteenth century they were imitated in the residences of wealthy Romans.[16]

When musical theater eventually opened to the public in Italian cities, financed for profit either by noble families or shareholders, it was confined to the six to ten weeks of Carnival—from December 26, the feast of Saint Stephen, to Shrove Tuesday, the day before the beginning of Lent. (In the eighteenth century, the season was

Teatro Farnese, Parma, Italy, 1628. Plan.

Teatro SS. Giovanni e Paolo, Venice, Italy, remodeled 1654. Plan.

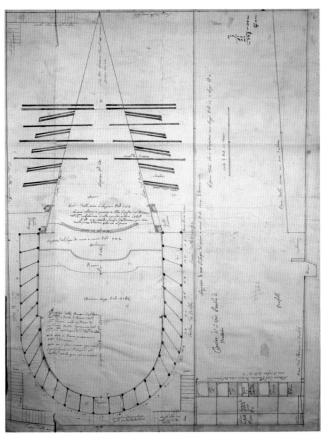

extended to the period of Ascension, to some of the church's autumn celebrations, and to festivities such as weddings and birthdays established by local nobility.)[17]

Residenztheater, Munich, Germany, 1753.

Venice was the seat of Europe's first public opera house, the Teatro di San Cassiano (1637). For almost the entire decade of the 1630s four such houses operated there, by 1660 offering as many as ninety different operas within a season.[18] By 1700 more than forty Italian towns had public opera houses.[19] Carlo Fontana, an apprentice of Gian Lorenzo Bernini who became prominent in his own right, rose to the challenge of the genre's extraordinary popularity.

For the paying public—the upper and professional classes who were replacing court spectators—Fontana renovated the Teatro SS. Giovanni e Paolo in Venice (1654): his refinement of the horseshoe produced a lasting model for opera house interiors. Within the U shape, the parterre was flattened and the walls were stacked with boxes. While the latter brought the largest number of spectators close to the stage, the effort to increase attendance filled the parterre with people and musicians. These, in turn, prevented performances from flowing into this space, thereby lessening their participatory quality, a quality that was further diminished by withdrawal in the mid-nineteenth century of the forestage projecting beyond the proscenium.

Members of the upper echelons of society who occupied the boxes treated them as extensions of their homes, readily tolerating inferior acoustics (due to the small openings to the auditorium and more sound-absorptive materials within).[20] During the Baroque period, boxes in Italy were tightly compartmented, with private anterooms in which socializing, preparing and eating food, and even lovemaking were the norm. In France boxes and galleries were more open, curtailing the most intimate activities. In both Italy and France, gaming rooms helped to assure the economic stability of the major houses.

The prototype offered by Fontana—the horseshoe, together with round and oval variations on the form, and tiered boxes—was adopted for the private theaters built by monarchs in the eighteenth century, with preference given to grand neoclassical facades. A good example is the 523-seat Residenztheater (1753) in Munich, designed and built by François de Cuvilliés, court architect to the elector of Bavaria Maximilian Joseph III. This lavish Rococo jewel included machinery for flattening the raked auditorium floor to become a ballroom. (The theater was altered when it was rebuilt on a different site in the 1950s, although it retained much of the original decoration.) Similar multiuse options existed in other opera houses of the era. A neoclassical version of the sovereign theater is Louis XV's oval, 712-seat Royal Opera of Versailles (1770) by Ange-Jacques Gabriel. Favoring spatial symmetry over decorative detail, its complex vertical sequence of galleries culminates in an adaptation of the Teatro Olimpico's grand colonnade.[21]

The horseshoe was also used for public theaters like La Scala in Milan (1778) by Giuseppe Piermarini, the largest—with 2,800 seats—and grandest of all opera houses when it was completed. The U shape and tiers of boxes have remained the favorite for opera seating for many reasons. Richard Pilbrow, the founder of Theatre Projects Consultants, traces the form back to early storytellers and performers around whom people gathered in a semicircle and then a full circle—West African griot and American Indian entertainers, for example. The pattern was developed in Elizabethan England at the Globe Theatre and imitated around the world by similar "courtyard"

Palais des Papes, Avignon, France, 1355. Courtyard.

Royal Opera, Versailles, France, 1770. Mid–nineteenth century watercolor, Eugene Louis Lami.

theaters. Within the Palais des Papes in Avignon (1355) is a courtyard in which stadium seating rises from a forestage jutting into the audience. In the twentieth century, technical progress made possible by iron structures and balcony cantilevers allowed alternatives to earlier raked stages and horizontal balconies, such as ski-slope balconies and stepped side boxes. Pilbrow is convinced that the horseshoe is the most efficient way to meld actor and public in an intimate relationship.[22] In addition, it boasts a good track record for the acoustics of opera and musical theater.

The horseshoe had another advantage. It allowed boxholders to look back from the stage to other members of the audience, an attraction that vied in importance with the actual performance until well into the nineteenth century. The royal box at the center of the horseshoe's first balcony was clearly visible to more members of the audience than the stage itself.

Nowhere was the horseshoe used to better effect at the time than at Charles Garnier's Paris Opera (1875), known as the Palais Garnier, whose architect was a young unknown when he won the commission. Part of Baron Haussmann's urban redesign of the French capital for Napoleon III, the opera had a commanding exterior that provided a spectacular terminus to one of the city's newly created boulevards. Its majestic public presence reflected the institution's role as a symbol of national glory after the trials of the Franco-Prussian War and became a focal point for the city's social life. With the splendor of its grand stairway, ample foyers, palatial promenades, and expansive backstage, the 1,979-seat Palais Garnier set a new world standard, outdoing even La Scala. The auditorium is equally successful for both opera and ballet (with which many opera houses share their stages), although sight lines and acoustics vary in quality. The fact that the areas designed to encourage the lively spectacle

La Scala, Milan, Italy, 1778. Exterior view, mid–nineteenth century engraving; interior view, aquatint, Antonio Bramante.

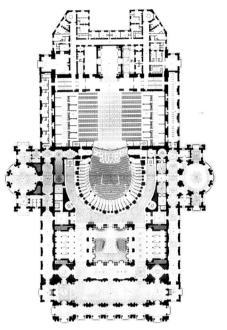

provided by the operagoers themselves take up more space than the auditorium clearly illustrates the priorities of the times. How these priorities have changed since the turn of the twentieth century is demonstrated by the gradual enlargement of backstage areas, necessitated by the change from scenery flats to mobile sets that characterizes modern theaters, with the Palais Garnier's counterpart, the Bastille Opera (Carlos Ott, 1989), leading the way in unprecedented size.

Palais Garnier, Paris, France, 1875. Exterior view; interior view, lithograph, *Le Monde illustré,* 1875; plan.

Innovations in Opera House Design: From Profane to Sacred

While Garnier's concept of public buildings, and especially theaters, as settings for social interplay remains valid, his hierarchy of circulation spaces and seating according to the audience's affluence and social standing was rooted in the past. Within a year of the Paris Opera's inauguration, Richard Wagner's iconoclastic Bayreuth Festspielhaus challenged these presumptions.

Wagner, who dismissed the commercial repertory system of his time as mindless, wanted to transform opera into the spiritually uplifting experience of theater in ancient Greece, and he was convinced that a new kind of building was needed to achieve this. As a conductor from 1842 to 1849 at the Dresden Opera House (1841), designed by Gottfried Semper, Wagner had ample time to discuss his project with the renowned architect. (At Dresden Semper expressed for the first time on the exterior the auditorium's cylindrical form, as was done subsequently at Bayreuth.)

The scheme for the Festspielhaus was financed by "mad" King Ludwig II of Bavaria, to whose dreams of grandeur the ambitious scheme appealed. Three architects worked on the undertaking, planned first for Munich but then moved to Bayreuth. Semper, the first designer, was eventually replaced by the younger Otto Brückwald in consultation with the technician Carl Brandt. But it was Wagner's concept that shaped the building, completed in 1876.

Wagner envisaged the new theater as the antithesis of the Markgräfliches Opernhaus (1748) in Bayreuth, an ornate extravaganza commissioned by Princess Wilhelmine, sister of Frederick the Great, from Giuseppe Galli-Bibiena, a member of the family that for over a century invented increasingly complex stagecraft for theaters throughout Europe. Because the composer conceived his opera house as a temporary structure that would be destroyed (like Valhalla) after the *Ring* tetralogy's initial performances, it was constructed with minimal means. (The Festspielhaus cost 428,000 marks compared with the Paris Opera's 28 million marks.)[23] An exposed timber frame (largely replaced with concrete in a 1962–74 reconstruction) with brick infill contains the auditorium, decorated in dull brown tones and without the usual embellishments. A stretched canvas glued to the flat wooden ceiling is decorated to evoke the canopy of an outdoor Roman theater. There were no foyers or any other amenities (originally, not even toilets) at this shrine to opera, whose very name (Festival House) evoked ancient Olympic celebrations with religious connotations. As revolutionary as its architecture was the Festspielhaus's location on a hill outside a small town, distanced from the temptations of a metropolis. It was to be a pilgrimage site removed from daily cares.

Even though Wagner rejected Semper's initial grandiose design for King Ludwig II's opera house in Munich (1865), aspects of that project informed the more modest house that was built (as well as three of Semper's other theater designs).[24] As with the scheme for Munich, the auditorium looked to antiquity: its layout is, in fact, remarkably similar to that of the Odeum of Agrippa built nearly two thousand years earlier (but unknown to the composer and his architects since it was excavated long after their lifetimes). The similar sight lines and the respective acoustics have proven excellent.[25] Both medium-size theaters work for what they were designed for: Wagnerian operas at the Festspielhaus, speech and small musical ensembles at the Odeum of Agrippa.[26]

The Festspielhaus's steeply raked fan shape, with aisleless rows of seats (known as continental seating) and two relatively small galleries at the back, broke with centuries of balconied, horseshoe configurations divided by aisles at the parterre level. Two proscenium arches, one in front of the other, with the space in between left in

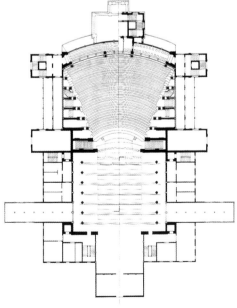

darkness, make the singers look larger than life. The illusion is reinforced by six parallel pairs of receding cross-walls at each side that hide doors to the vestibules and mask the disparity between the hall and its square container. To avoid any distraction from the orchestra, the pit, sunken below the stage, is covered by a wood and leather hood on the audience side to make what Wagner called the "mystical abyss": a device emphatically separating theater (the stage) from life (the audience).[27]

By dampening the high-frequency sounds emanating from the strings and woodwinds, the "abyss" emphasizes midrange and low frequencies. And because bass tones are omnidirectional, the arrangement produces the illusion of music enveloping the audience. Acoustical and visual effects, including the first dimming of light in a theater, contributed to Wagner's goal of producing a total work of art, a *Gesamtkunstwerk*, for which every aspect of the performance was part of a seamless whole.[28]

Wagner's new kind of democratic seating offered, for the first time since the theaters of ancient Greece and Palladio's Teatro Olimpico, rational vertical and horizontal sight lines to all attendees. Paradoxically, the egalitarian seating served an elitist vision of how opera should be performed. The Bayreuth festival hall anticipated in terms of operagoing what twentieth-century architects would achieve for viewing modern art when they designed pared-down white cube galleries in which art is revered and nothing detracts from insulated displays.

With his Festspielhaus, the composer created the ideal opera house for the *Ring*, and with *Parsifal* (first performed in 1882), the ideal piece to exploit its particular acoustics, much in the same way that earlier composers wrote music to take advantage of the particular characteristics of a cathedral, church, or hall. In fact, Wagner's widow Cosima insisted that *Parsifal* be performed nowhere else, a ban unbroken until the Metropolitan Opera's 1903 production. Because the theater is designed so specifically with Wagner's late operas in mind, some critics cite a loss of intensity and detail for other music, even for the composer's earlier operas.[29]

Furthermore, not every conductor accepts the dark, covered pit with its six descending levels for the musicians as willingly as the conductor and pianist Daniel Barenboim, who dismisses complaints about conducting at the Festspielhaus and praises its unusual acoustical quality. Instead of sending sound directly to the audience, mixing it with sound from the stage, the sunken pit projects the orchestra's music to the stage; only after combining with the singers' voices does it rebound to the audience. Barenboim compares the demands of conducting at Bayreuth to those of scuba diving, a favorite sport of his: "When you deal with a problem underwater," he says, "you have always to go slowly, with rounded gestures, and anticipate what is coming. Otherwise you are in trouble."[30]

The uninterrupted fan-shaped seating initiated by the Festspielhaus was adopted as early as 1889 by Adler & Sullivan in the magnificent Auditorium Theatre in

Chicago; it was popular from the 1920s to the early 1970s for concert halls in the United States and in the 1920s and 1930s for movie houses in the United States and Europe. The 4,200-seat Chicago hall was part of what was, at the time of its completion, the largest building in the country (including offices and a hotel) and the costliest ($3.2 million) in Chicago. Despite the auditorium's reputedly excellent acoustics, the fan shape was generally dropped when it became apparent in similar theaters constructed afterward that by reflecting sound toward the back corners of a concert hall, the splayed side walls delayed reflections to the center seats. Acoustics were also affected by increased size: what worked well for an audience of 1,800 at Bayreuth was not valid for later auditoriums that seated many more.

The Concert Hall

The genealogy of concert halls is different from that of opera houses. The latter derive from ancient Rome—from semicircular plans, as reinterpreted in the Renaissance, and from fan-shaped seating, as adopted by Wagner. Conversely, until the middle of the twentieth century, the vast majority of purpose-built concert halls consisted of long, narrow, rectangular rooms. Such rooms started to appear in London late in the seventeenth century to accommodate the orchestral concert works being developed, in competition with church and court music, from opera overtures and dance suites. From the taverns used for performances during the Commonwealth period (1649-60), when theaters were closed by Charles I's Puritan anti-playgoing ordinance, and from the private homes that were opened to the public, musicians moved to large rooms being built at approximately the same time as the first theaters—which were used primarily for opera.[31]

Music Meeting, a great room in York Buildings, a fashionable development of around 1675, was followed in the next century by numerous so-called music rooms. Among these was the approximately eight-hundred-seat Hanover Square Rooms in London (1775), designed by Giovanni Gallini, Johann Christian Bach (a son of Johann Sebastian Bach), and Carl Friedrich Abel. Contributing to the fame of the Hanover Square Rooms, demolished in 1900, was Joseph Haydn's composition of his London Symphonies (Nos. 93-104) specifically for the hall, and his conducting of them there in the course of two visits between 1791 and 1795. In Edinburgh Robert Mylne used the Teatro Farnese in Parma as a model for St. Cecilia's Hall (completed in 1763, and from 1798 on subject to major changes). The music room, a five-hundred-seat oval (a favorite shape for early auditoriums), had tiered benches around the walls; the two halves of the audience faced each other across an empty space at the center, with the stage at one end.

Between about 1675 and 1775, while middle-class British audiences enjoyed paid, public concerts, music in Europe remained exclusively a private affair, in the salons of the well-to-do and the nobility, or in the large banquet halls and horseback-riding

Hanover Square Rooms, London, England, 1775. Lithograph, *Illustrated London News*, 1843.

Haydnsaal, Esterházy Palace, Eisenstadt, Austria, 1672.

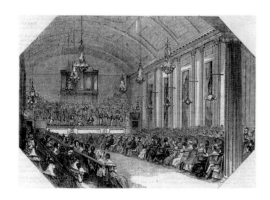

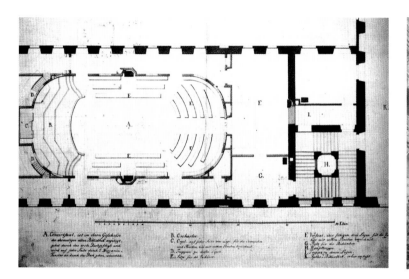

arenas of ruling prelates. Typical of these rectangular rooms with a raised platform at one end and often a balcony on three sides is the four-hundred-seat Haydnsaal in Esterházy Palace in Eisenstadt, Austria, located about fifty miles outside Vienna. A medieval fortress, the palace was renovated between 1663 and 1672 according to the plans of the Italian architect Carlo Martino Carlone, with the help of Sebastiano Bartoletto. The residence, including the concert hall, features simple painted imitations of the elaborate three-dimensional decor that graced comparable city abodes. One of Haydn's first positions as an ambitious young composer and conductor was assistant Kapellmeister (for four years beginning in 1761) at Esterházy Palace. His Symphonies Nos. 2–25 and 27–34 were composed for its concert hall, scrupulously taking into consideration the long reverberation time.[32]

Not until the mid-eighteenth century did small (four hundred seats) public concert halls begin to replace the coffeehouses in which public concerts had taken place in Europe. (Johann Sebastian Bach's weekly Collegium Musicum concerts at Zimmermann's Coffee House in Leipzig are a noteworthy instance of the custom.) These auditoriums more or less replicated the court halls' shoebox shape. A good example was the destroyed Altes Gewandhaus in Leipzig, Germany (1781), by the architect J. C. F. Dauthe. A rectangular room with curved corners, it featured a flat ceiling with coved edges and upper-level seating lining the full length of the sides of the auditorium, with a raised box on each flank of the stage and on the wall opposite the stage. As at St. Cecilia's Hall, the two parts of the audience faced each other across an unoccupied central aisle, playing up the importance of social interaction. The Leipzig auditorium's insertion into part of the then existing Garment Traders Hall followed

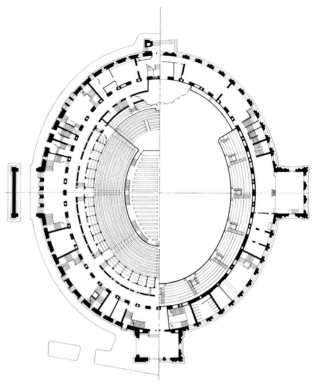

Royal Albert Hall, London, England, 1871. Interior view; plan; exterior view.

the Renaissance tradition, as at the Teatro Olimpico and the Teatro Farnese, of constructing a timber theater within a brick building.[33]

The acoustics of these spaces, which were enhanced by early reflections from the parallel side walls (providing clarity) and later reflections from the back of the hall and by reverberation (multiple reflections from all directions enveloping the listener),[34] established a more influential format than the various designs and varying acoustics of British concert halls. The shoebox was again the model for the nineteenth-century boom in larger auditoriums that responded to the public's growing hunger for music. This was when concerts of classical music could assume mammoth proportions, requiring exceptional containers. In England and the United States at what were referred to at the time as Monster Concerts, as many as five hundred musicians might perform for audiences nearing ninety thousand.[35] The concerts Hector Berlioz organized in London's Crystal Palace at the first world's fair (the Great Exhibition of 1851) and at the Machinery Hall at the Exhibition of Industrial Products in Paris (1855), as well as Gustav Mahler's Symphony No. 8 (known popularly as the Symphony of a Thousand) in the Neue Musik-Festhalle at Munich's International Exhibition grounds (1906), represent this development.

A notable nineteenth-century exception to the shoebox was the Royal Albert Hall in London (1871), in which the oval auditorium achieved a colossal scale. Inspired by the Roman arenas at Arles and Nîmes, the architect Sir Henry Cole and Captain Francis Fowke of the Royal Engineers (who was replaced after his death by his colleague Colonel Henry Darracott Scott) substituted red brick for stone and interrupted the facade by arched porches for carriage arrivals and departures at the north, west, and

Schauspielhaus, Berlin, Germany, 1821. View of concert hall, 1920; exterior view; plan, with concert hall at left.

east. The architecture's neoclassical aspirations were emphasized by a terra-cotta frieze with allegorical and historical scenes and figures, in addition to an inscription acknowledging Prince Albert's inspiration. The prefabricated double-skinned, iron-and-glass roof dome was a daring feat of engineering at the time.

Initially conceived as a thirty-thousand-seat amphitheater, the project was of a magnitude to match Queen Victoria's wish to properly honor the memory of her husband, Prince Albert, with a building that fulfilled his own plan for a hall of arts and sciences. The size was soon reduced to a more practical seven thousand seats (and, after a renovation completed in 2003, 5,250), with three tiers of boxes, above which are balcony seats and a gallery for standing or promenading. This huge theater followed the lead of the municipal buildings that had begun to appear by mid-nineteenth century in northern cities like Birmingham and Liverpool, where upwards of five thousand people attended concerts.

Even before the Royal Albert Hall's official opening, tests revealed multiple acoustical defects, among them echoes and blind spots. It is intriguing that despite the failure of a series of efforts to correct these problems, the hall has been tremendously successful throughout its history, typically with a million visitors a year.[36] The gallery and the large arena (where standees may buy cheaper tickets) are extremely popular, especially for the well-known Promenade Concerts that have been offered there every summer since 1941.

By 1900 the classical concerts of the hall's early years were alternating with wrestling and boxing matches; in later decades popular concerts by the Beatles, Rolling Stones, and other megastars were featured. In the last third of the twentieth century, high and low culture continued to rub shoulders in the hall with the introduction of opera, ballet, sumo wrestling, and ice skating.

Well before the Monster Concerts were taking place in oversized exhibition spaces, a new generation of purpose-built concert halls—larger than those of the eighteenth century but considerably smaller than the exhibition pavilions—had begun to emulate Karl Friedrich Schinkel's Schauspielhaus auditorium in Berlin (1821). The neoclassical architect placed the concert hall in the west wing of his "play house," a splendid structure whose grand Ionic portico and broad entrance stairway dramatize its relationship with the neighboring twin French and German churches to make one of the city's most beautiful squares. Entered via three doors in the rusticated podium, the concert hall, seating up to 1,677, had a single balcony, wall sculptures, and gold ornamentation.[37] Its conventional shoebox shape contrasted with the innovative drama theater, which was inspired by the Greek semicircular plan.

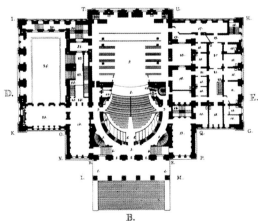

Of the many nineteenth-century halls that looked to the Altes Gewandhaus and the Schauspielhaus concert hall, two outstanding representatives are, respectively, the Musikverein in Vienna (1870) by Theophil von Hansen and the Concertgebouw in Amsterdam (1886) by Adolf Leonard van Gendt. The Musikverein was founded in 1812 by a society of amateur musicians typical of the era (they were replaced in 1842 by the Vienna Philharmonic). By 1870 the Musikverein had inaugurated its own concert hall, built, as was the Paris Opera in 1875, during a period of massive urban reconfiguration. The Danish-born architect von Hansen, a devoted neoclassicist who had studied and worked in Athens, suited so perfectly the historicizing taste of the time that in addition to the Musikverein, he was responsible for four of the most important public buildings (and a private palace) along the Ringstrasse, the city's new major artery.

The Musikverein's pedimented, polychrome facade, with its elegant adornment of statues, is an apt introduction to the extravagant 1,680-seat Great Hall, the larger of the two auditoriums (three small ones were added in 2004). Gilt is used so liberally in this Grosser Musikvereinssaal—for the pairs of caryatids lined up below the cantilevered balcony, for the decoration of the balustrade extending to three sides of the room, and for the ceiling—that the room is commonly referred to as the Golden Hall.

From its first concert, the Golden Hall had an equally golden sound. Many factors have been credited for this success: the shoebox shape, the wooden ceiling that

Musikverein, Vienna, Austria, 1870. Exterior view; Golden Hall; parterre plan.

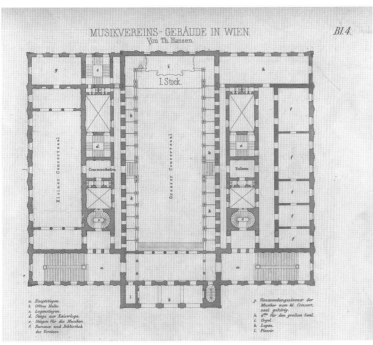

Concertgebouw, Amsterdam, Netherlands, 1886. Exterior view; interior view; plan.

hangs from beams, the more than forty high windows, and the extensive articulation of an interior constructed mainly of plaster on brick or wood lath. Also mentioned are the hall's relatively small size and its tightly spaced wooden seats, smaller than today's since audience members were then smaller. (The Musikverein gives just 315 cubic feet to each member of the audience, compared with, for instance, nearly 410 cubic feet at the 1989 Meyerson Symphony Center in Dallas.) Nonetheless, the key to the Musikverein's excellent acoustics has remained elusive.

In Amsterdam, the Concertgebouw's Main Hall (there is a small, 478-seat chamber music hall as well) also has excellent acoustics. But compared to Vienna, the Amsterdam venue had a slow start: eleven years of architectural alterations to correct the acoustics were needed after the building's completion.

Within the Main Hall, which seats 2,000, steep stadium choral seating for 311 singers (occupied by the public when not used for a chorus or additional musicians) rises at either side of the organ behind the flat stage.[38] This stepped arrangement provides a dramatic entrance for the musicians who come in at the top and descend to the stage. As with seating in the St. Cecilia and Altes Gewandhaus halls, among others, the arrangement is another move toward the surround seating that would be developed in the middle of the twentieth century. A slender balcony supported by thin Corinthian columns edges three sides of the hall; on opposite sides, immediately under the ceiling, are rows of semicircular windows (so-called thermal windows divided into three panes by two mullions). A structural renovation and addition of a new entrance wing (1988) by Pi de Bruijn mercifully left the large and small halls intact.

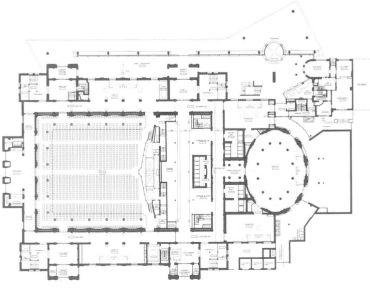

The Concertgebouw's marshy site, then at the city's edge, was chosen by Pierre Cuypers, the architect of the nearby Flemish Renaissance Rijksmuseum. Van Gendt had apparently tried to relate the concert hall facade to the museum, with the strange addition of a pedimented portico. The Concertgebouw has long since been engulfed by the city, whose heavy traffic and streetcar lines form a new kind of hurdle in reaching it.

Unlike contemporary auditoriums, the Vienna hall and other similar historic spaces for music used no professional acoustical consultants. Although by the mid-nineteenth century a scientific understanding of acoustics existed, it was Wallace Clement Sabine, applying his newly discovered techniques to McKim, Mead & White's Symphony Hall in Boston (1900), who launched the specialty.

Many recent concert halls continue to imitate these nineteenth-century auditoriums, just as recent opera houses remain wedded to the more than three-hundred-year-old symmetrical horseshoe configuration. The music critic of the *New Yorker,* Alex Ross, posits, however, "Music shapes the concert hall and vice versa. The new styles and new timbres of the twentieth century require different kinds of spaces."[39]

The Democratization of the Concert Hall and the Berlin Philharmonie

The sacralization of culture, by which the arts lost their representational and recreational functions and became something to be worshipped in their own right, began in the middle of the eighteenth century, thanks largely to the venerational aesthetic of the German archaeologist Johann Joachim Winckelmann.[40] The attitude developed in the nineteenth century: in addition to being a time of great music, the era also saw the flowering of music journalism, criticism, scholarship, and historiography. Such novelties, paired with the rise of music education and music publishing, plus the ubiquitousness of pianos in middle-class homes, made people more passionate about, and respectful of, music itself, as opposed to music as a social event.[41] Indeed, the conductor controlling the orchestra and audience was like a priest handing down a commandment from on high. What was gained thereby in focus was offset by the suppression of the audience's former responsiveness. Schinkel, for one, known for his masterful theater designs, articulated early in the nineteenth century the importance of reciprocity between public and players, a concept that was taken up in the twentieth century by theater professionals at the Bauhaus and is still in effect.[42]

Just as modern visual artists devised ways to overcome a passive viewing of their work, so forward-looking music professionals have welcomed novel ways of presenting classic repertoire as well as new architecture to accommodate experimentation.

David Robertson, music director of the St. Louis Symphony Orchestra, says, "It is more difficult to perform new music in a classic hall or opera house because they raise expectations of something familiar."[43] A dedicated proponent of contemporary music, Robertson asserts that "the moment you go into an unusual space, unusual music doesn't sound so unusual." Daniel Barenboim echoes the thought: "The classic shoebox has too many associations with the past."[44]

Since the first quarter of the twentieth century, there have been many experiments in the design of sites for sound: some have been inserted into renovated historic or industrial spaces, others are newly invented typologies. In 1920, the inaugural performance of the Salzburg Festival—Hugo von Hofmannsthal's *Jedermann* ("Everyman") in a production by the stage director Max Reinhardt on the steps of the Cathedral Square—marked a major departure from standard theater presentations. In 1926, the Summer Riding School in Salzburg—three tiers of shallow arcades hewn by Johann Fischer von Erlach into the walls of a disused quarry in 1693—was renovated by the architect Clemens Holzmeister for concerts and opera. It is still in use, the hard rock face providing an exceptionally effective reflective surface not unlike that of Greek and Roman skenai. In the late 1940s, new festivals in the Palais des Papes at Avignon and in the medieval Théâtre de l'Archevéché at Aix-en-Provence continued the Salzburg example of using old sites for the performance of music, paving the way for France's great era of festivals in the 1980s. Abandoned buildings, including a cereal trading market in Toulouse and a former gasworks in Amsterdam, have joined these historic buildings. Between 2000 and 2010, led by Germany's large-scale conversions in the towns of Bochum, Duisburg, and Essen, effected by Gerard Mortier in 2002 for the Ruhrtriennale, converted industrial spaces have offered more neutral alternatives to historic settings.

In the United States in the 1930s, musicians were performing new music in numerous small, unconventional spaces. Evenings on the Roof (of a private home) in Los Angeles was founded in 1939 and is still operating. (In 1954, the program was renamed the Monday Evening Concerts; since 2006, they have been held in Walt Disney Concert Hall's REDCAT auditorium and at Zipper Hall, a 435-seat auditorium on the Colburn School campus.) In the 1960s, as theater directors in Europe and eventually Japan came to prefer converted religious, industrial, and other unconventional spaces, in the United States even classic programs were being staged in less formal conditions. In an attempt to capture waning audiences, the National Endowment for the Arts adopted venues reminiscent of the earlier, iconoclastic preferences of the Works Progress Administration, staging concerts in non-theaters indoors, outdoors in parks, and in at least one case in a shopping mall.[45]

If Wagner's music and the way it was performed in the Festspielhaus were the pinnacle of culture's sacralization,[46] Hans Scharoun's Berlin Philharmonie (1963) initiated an approach to concert hall design that reflected the changed spirit of this more recent era. Scharoun was a largely self-taught Expressionist architect with a few notable accomplishments in the 1920s and 1930s. But until the Berlin project, his numerous competition-winning designs had remained unbuilt. It was perhaps the unique state of the post–World War II city that allowed him, at the age of seventy, to realize for the first time one of his successful entries.

West Berlin at the time was a ruined and politically fragmented city. The government's siting of this cutting-edge concert hall within view of the divisive Soviet wall was a bravura statement, and Scharoun's innovative bringing together of performers and public was a welcome unifying gesture. The Philharmonie was the first in a series of buildings in the Kulturforum that now includes Mies van der Rohe's Staatliche Museen (1968); the Staatsbibliothek (Scharoun and Edgar Wisniewski, 1978); and the Gemäldegalerie (Hilmer & Sattler, 1997).

Supported by the conductor Herbert von Karajan, the architect and the acoustician, Lothar Cremer, broke down old barriers, abandoning the sacred aura of a templelike exterior and an altarlike stage in a shoebox space. Drama theaters in the round had been built in the twentieth century since the late 1940s, but the Philharmonie was the first modern adaptation of that configuration to a concert hall.

Berlin Philharmonie, Germany, 1963. Exterior view; interior view; plan; section.

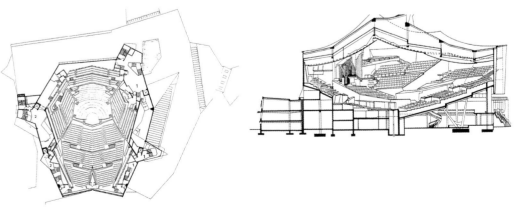

Scharoun designed the Berlin auditorium so that the many vertical planes of the seating terraces and the tentlike ceiling (almost sixty-six feet high above the conductor's podium) compensate for the lack of side walls to reflect sound. The design illustrates why all shoebox halls—thought to be a guarantee of acoustical quality—are not equally satisfactory: acoustics are based on reflecting patterns, not on Euclidean geometry.[47] Acoustical perceptions are also influenced by the listener's other senses—a phenomenon known as psychoacoustics. For visual and aural art context influences perception. In the same way that the effect of an artwork depends on lighting, wall and floor color and texture, and the relationship to nearby objects, so listeners are influenced by their surroundings when they hear music.[48] The Berlin Philharmonie is one demonstration of this: its fine acoustical reputation owes a great deal to the visual pleasure experienced in its seductive space, without which it would not receive such high marks. Yasuhisa Toyota, for one, the acoustician for Disney Hall and many other highly thought of auditoriums, says of the Berlin auditorium, "The acoustics are not bad, but they cannot be judged with musicians other than those of the excellent Philharmonic Orchestra, which has overcome the hall."[49]

The Main Hall and the smaller Kammermusiksaal (the latter encouraged by the composer Luigi Nono[50] and completed in a poor approximation of Scharoun's style by Wisniewski in 1987 after the older architect's death) returned to an earlier concept of encircling folk musicians. Scharoun saw the space as a valley with the orchestra at the bottom (but not at the exact center) and vineyards climbing the surrounding hills. Arrangement of the wood-paneled hall's 2,218 seats in terraced circular segments in front of, behind, and on either side of the orchestra achieved an unprecedented feeling of egalitarianism and intimacy, with no listener more than one hundred feet from the stage. The orchestra pit could be raised and lowered hydraulically (it was replaced in 1975 by forty movable pedestals), and provision was made for a variety of theatrical lighting arrangements.

Three small platforms distributed around the top part of the auditorium and one additional space acknowledge the occasional need to situate musicians away from the orchestra. Customs die hard, however. These platforms have been used for nineteenth- and occasionally twentieth-century music that calls for spatial placement, but rarely for new compositions, and some conductors who have performed there are unaware of their existence.

A Piranesian series of levels and stairways in the foyer was as unusual at the time of completion as the concert hall itself. The modesty of exposed-concrete surfaces and painted-metal handrails matched the democratizing intention of the auditorium. Reflecting the hall's shape, the foyer's sculptural ceiling is a structural support for the Main Hall above. The interpenetrating interiors, bathed in daylight and colored here and there by tinted glass, were conceived by the architect as a lively cityscape that would contrast with the serene landscape of the auditorium. Passage of the public through the foyer's complex spaces, in which basic amenities, food, and drink are available, and socializing occurs, is a spectacle in its own right. The design is as sensitive to the needs of today's audiences as the Palais Garnier was to those of its Second Empire public.

The building's asymmetrical exterior, possibly influenced by Le Corbusier's pilgrimage church at Ronchamp (1954), expresses its interior volumes. In the 1980s, in accordance with Scharoun's wish for a dazzling effect, it was clad in yellowish aluminum and polyester panels. The innovations introduced by the Berlin Philharmonie—the vineyard auditorium arrangement, a different kind of circulation space—soon appeared in other concert halls and are still influencing new concert halls throughout the world.[51]

Music to Serve the Space

The addition of electronic and experimental music to the classic contemporary repertory has occasionally revived the tradition of music composed for a specific space, most often for a temporary structure. One such piece was Edgard Varèse's *Poème électronique* for Le Corbusier and Iannis Xenakis's hyperbolic paraboloid pavilion, which was constructed with a series of cables overlaid with a skin of concrete panels. It was commissioned by the Philips Radio Corporation for the 1958 World's Fair in Brussels. Le Corbusier described the round enclosure as a "stomach" that was to digest and then disgorge the crowds. Xenakis, the brilliant architect-composer employed by Le Corbusier at the time of the commission, worked from the master's schematic layout. The interior was acoustically dampened to assure the clarity of the complex electronic sounds and visually darkened to sharpen the projections Le Corbusier designed. Instead of images that would promote the client's products, the projections and everything else about the pavilion, including its structure, illustrated the evolution of technology.

Xenakis used curvilinear forms for the walls and roof to create a seamless whole in which approximately four hundred loudspeakers played real and manipulated sounds. Varèse's piece, on three-track tape, used sources such as percussion, electronic tones, machines, and the human voice. Images that included animals, religious objects, art, architecture, film, and war scenes, paired with colored lighting effects, were projected on the walls. The *Poème électronique* and Xenakis's own tape composition, *Concret PH (Paraboloïdes Hyperboliques)*—a pun on concrete, the material, and *musique concrète*, the technique—were constantly sent around to different loudspeakers. Constructed entirely from the crackling sounds of burning charcoal, *Concret PH*

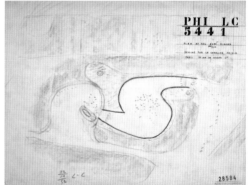

Philips Pavilion, Brussels, Belgium, 1958 World's Fair. Exterior view; interior view; plan.

was heard before and after the show, as the public entered and exited. The machine-driven multimedia event, starting before civilization and ending with the question of humanity's survival, was repeated several times a day in an immersive environment—another form of *Gesamtkunstwerk*—that attracted over a million visitors during the six months of the fair.[52]

A one-on-one collaboration between musician and designer also took place between the composer Karlheinz Stockhausen and the architect Fritz Bornemann, aided by the artist Otto Piene, for the West German Pavilion at the 1970 Osaka World's Fair. The intention was to reverse the idea of Scharoun's Berlin Philharmonie, where the audience is seated around the orchestra. In the Osaka pavilion, it was the audience, sitting in a concentric arrangement on a circular, sound-permeable grating at about mid-level in a spherical concert hall, that was surrounded by fifty groups of loudspeakers arrayed in three dimensions.[53] Soloists played or sang either on a podium at the side or on one of six balconies built above the audience platform.[54] A wide repertoire, mostly by Stockhausen, included various permutations of natural acoustic and/or electronic music on multitrack tape, electronic equipment like shortwave radios, with or without one to nineteen live vocal or instrumental performers. As at the Philips Pavilion, listeners were enveloped both within the space and by the sound, but in contrast to the success of the Corbusier/Xenakis installation, the Stockhausen/Bornemann experiment rarely had more than thirty people in its three-hundred-seat space.[55]

Renzo Piano performed a similar experiment for a work by Luigi Nono, who inventively fuses old and new, live and electronic music. For Nono's otherworldly, spatial opera *Prometeo: La Tragedia dell'ascolto (Prometheus: The Tragedy of What Is Heard)*, created for the 1984 Venice Biennale, Piano designed a wooden "ark." (The

West German Pavilion, Osaka, Japan, 1970 World's Fair. Interior view; exterior view; section; plan.

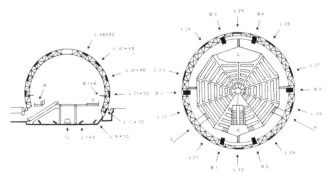

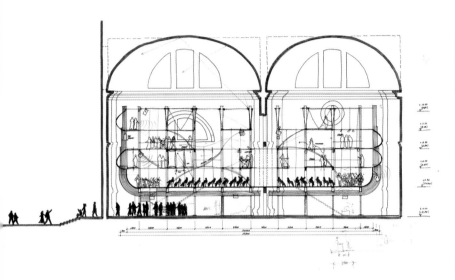

opera's subtitle, *La Tragedia dell'ascolto*, conveys Nono's rejection of the usual subordination of sound to sight.) Unlike earlier freestanding pavilions, Piano's was a modular structure erected first inside the Church of San Lorenzo and then, the following year, in the unused Ansaldo factory in Milan. (In one of many subsequent performances, *Prometeo* was given in 1998 at the Akiyoshidai International Art Village, where a concert hall, designed especially for this piece and other multifocal, spatially oriented compositions, was completed that year by Arata Isozaki.)[56]

Piano's response to his friend Nono's request to "make a symphony with me"[57] was a multilevel structure built in prefabricated laminated timber, fireproof plywood cladding panels, and tubular steel, the whole resembling the skeleton of a ship's hull. The installation positioned the four-hundred-person audience at the center of the lowest level, with the eighty members of the chorus, soloists, readers, and percussionists moving through and around them on three upper levels of walkways and stairs, following the two conductors on video screens. The architecture thus allowed Nono to break what he regarded as the stranglehold of unidirectional sound; instead, fragmented sounds, modulated by computer and amplified, come together with unique effects and descend vertically in a three-dimensional mixture heard differently by each member of the audience. Among the many precedents for his position the composer cited the separation of the two choirs at San Marco in Venice (see Introduction).[58]

Instances of musical spatialization go back to antiquity. In the first century CE, Seneca described an auditorium surrounded by brass instruments in which the aisles were filled with singers.[59] Others who composed music with spatially separated performers include Monteverdi in the seventeenth century; Bach and Mozart in the eighteenth century; Beethoven, Hector Berlioz, and Wagner in the nineteenth century; and Gustav Mahler, Charles Ives, John Cage, Iannis Xenakis, Karlheinz Stockhausen, Pierre Boulez, Luciano Berio, Henry Brant, Benedict Mason, Wolfgang Rihm, and Georg Friedrich Haas in the twentieth and twenty-first centuries, with the number of such composers increasing in recent decades. The conductor and composer Michael Tilson Thomas says, "Placing instruments offstage widens our acoustical perspective."[60] Even for nonspatial music, the choreographer Merce Cunningham said, "I like to bathe the audience in sound, surrounding them, instead of the way they hear it in an ordinary hall."[61]

Spaces to Serve the Music

Modular theaters, consisting of movable floor, and possibly ceiling, components that can transform a space from proscenium to thrust to arena, offer a practical alternative to interiors devised for a specific orchestra, type of music, or composition. They elaborate on the concept of Walter Gropius's unbuilt two-thousand-seat Total Theater project of 1927, which envisaged reconfigurations that would establish different relationships between the stage and the audience for a variety of genres including concerts and opera. Gropius conceived the project for productions at the Bauhaus Dessau by the director Erwin Piscator, originator of the Proletarian Theater in Berlin and later of the New School Dramatic Workshop and Studio Theater in Manhattan.

As the founder of the Bauhaus, Gropius sought a union of all the arts that took still further Wagner's concept of the *Gesamtkunstwerk*. In the 1920s, visions of the total work of art that included life itself were seen as an antidote to the isolation of specialization.[62] Piscator's determination to eliminate the separation between spectator and performer made him an ideal client for the new theater typology designed by Gropius and his assistant Stefan Sebök, which the director envisaged as a flexible "theater machine."[63] Not only could the main stage be repositioned, but a peripheral stage encircled the audience under an aerial platform: spectators could be completely enveloped by the action. A back-projection screen above the peripheral stage was to show film complementary to the main attraction, which would be one of many types—including music, sports, parades, and exhibitions—that were to take place in the theater.[64]

Common for drama, modular theaters can also work well for the performance of music. Yasuhisa Toyota, for one, claims that modular theaters offer no specific acoustical problems.[65] These interiors, best-termed multiuse, are distinct from the multipurpose theaters that were introduced in the United States in the late 1950s for

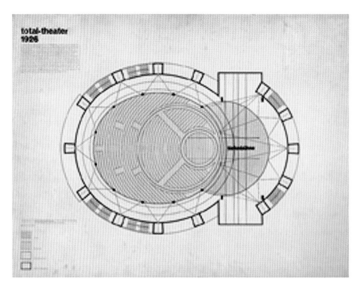

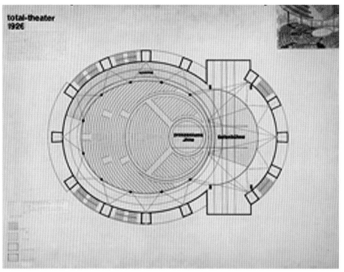

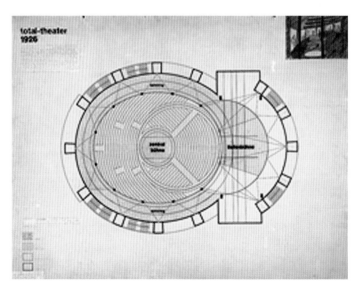

anything from classical opera to popular music and even sports events.[66] Usually created for at least 2,500 seats, the wide, fan-shaped rooms, many with multiple deep balconies, attempted to achieve acoustic variability by finding average criteria for speech and music, inevitably failing both. Heavy steel-contained shells created an unpleasantly hard, edgy sound, and only gradually have technical advances such as coupled tunable concert shells, curtains, and acoustical chambers (spaces that among other functions can be opened to the concert hall to alter reverberation time) produced satisfactory multiuse theaters.[67]

Despite the many small (around four hundred seats) modular rooms that now routinely supplement most new auditoriums, there is stiff resistance to reconfiguration, with the expense of labor usually cited as a deterrent. Whether or not to adopt a modular format for larger auditoriums is as controversial an issue among music professionals as the art world's division between those who favor neutral, rectilinear containers and those who opt for interactive, sculptural ones. But it is absurd to consider any of these options in the abstract, without regard to the kind of art or music involved.

"Flexible spaces are the only answer to help [opera] move forward," says George R. Steel, general manager and artistic director of the New York City Opera.[68] Representative of the opposite approach, for concerts, is Ara Guzelimian, provost and dean of the Juilliard School, who dismisses the need to reconfigure performance spaces. According to Guzelimian, "Most music is unidirectional. Spatial music is a rare exception, like Pierre Boulez's *Répons* [1981–84], for which soloists at six satellites placed at the peripheries of the audience complement the orchestra at the center."[69]

Répons is a good example of how bold new forms of music, barely conceivable in more standard conditions, may be inspired by a particular space.[70] Live electronic manipulation of sound by Boulez and his assistant Andrew Gerzso was facilitated by the advanced technology and computer science at Paris's Institut de Recherche et Coordination Acoustique/Musique (IRCAM, Renzo Piano and Richard Rogers, 1978, part of the Centre Culturel Georges Pompidou, expanded to the adjacent brick-clad tower by Piano in 1989). It was to participate in the creation of this facility that Boulez returned to his native France in the early 1970s after a prolonged absence.

Boulez's passionate commitment to innovative—even revolutionary—composition and new kinds of performance spaces spurred his involvement with the Espace de Projection, a room buried deep within IRCAM, where *Répons* was composed. At the time of construction, it was the holy of holies among modular theaters, carrying acoustical flexibility to an unprecedented extreme. The 46-foot-high walls consist of 172 triangular metal panels grouped in squares. Each side of the panels reacts differently to sound, and they can be adjusted to 370 alternative positions. Three movable sections of the ceiling and partitions on rolling beams modify the room's volume,

Total Theater project, 1927. Plans.

Espace de Projection,
IRCAM, Paris, France,
1978. Interior view.

Teatro Lirico, Cagliari,
Italy, 1965. Interior and
exterior model views.

and mobile bridges allow the placement of musicians, microphones, loudspeakers, and lighting throughout.

As is true for the Centre Pompidou, IRCAM owes its concept of flexibility largely to Cedric Price's unbuilt Fun Palace (1963–65), an insistently technological structural grid of steel-lattice columns and beams within which the suspended programmatic components—including theaters—could be assembled and disassembled as needed. Although it was intended for research only, the Espace de Projection now accommodates four hundred people for ten to twenty public concerts and dance performances annually.

An unbuilt competition entry (1965) for the Teatro Lirico of Cagliari, the capital of Sardinia, proposed a particularly ingenious version of the modular theater. It was to be constructed for the most part of standardized prefabricated metal parts. The Roman architect Maurizio Sacripanti, a high-tech modernist with an affinity for modular architecture, designed a system of electrically moved, prismatic floor and ceiling modules that run on racks joined to a supporting metal grid. The system could within twelve minutes transform the stage itself, the theater configuration, and the theater's size, both physically (from 3,000 to 1,848 or 1,152 seats) and acoustically, to complement what was to be performed in it. This would include, in addition to drama and music, a wide range of social functions, such as conferences and exhibitions. The exterior reflected the modular nature of the interior.[71]

Noteworthy among the venues built during this period of experimentation are the Walt Disney Modular Theater (1973), by Ladd & Kelsey with Jules Fisher Associates as theater consultant, for the California Institute of the Arts in Valencia, and the stage director Peter Stein's Schaubühne am Lehnines Platz (1981) in Berlin, by Jürgen

Sawade. The Schaubühne was erected inside the shell of a 1920s cinema by Eric Mendelsohn that had been largely destroyed during World War II.

The adjustable wall and floor panels of the CalArts facility offered greater physical flexibility than that enjoyed by many college theaters in the United States by this time. Those who know Woody Allen's 1973 *Sleeper* will be familiar with the fortress-like building, which figures prominently in the film. The four-foot-square modules that compose the theater's floor can be moved pneumatically to varying heights, and similar wall panels allow entrance from any point. A modular flying system for lights and scenery contributes to the space's adaptability. The Schaubühne's three performance spaces can be used separately or together, with spatial adjustments made possible by hydraulically manipulated floor components.

The Cagliari project and the Valencia and Berlin theaters were designed primarily for spoken drama. For the performance of music, it was not until the completion of Christian de Portzamparc's modular concert hall at the Cité de la Musique (1995) in Paris that Boulez was able to apply his ideas about spatial compositions to a public hall. The maestro is the first to admit, "Flexibility depends on size. For an audience of more than 2,000, you can't change that much; for a medium-size auditorium of 1,000 to 1,200, you can do more. And an auditorium of 400 to 600 is easily changed."[72]

The thousand-seat concert hall of the Cité de la Musique, which incorporates the National Music Conservatory, is set within a wedge-shaped building located in a working-class neighborhood of northeastern Paris. A *grand projet* of President François Mitterrand, the complex stands between the heavily trafficked Avenue Jean-Jaurès and the Parc de la Villette. Once inside the Cité, theatergoers reach the auditorium by

Walt Disney Modular Theater, CalArts, Valencia, California, 1973. Interior view; plan.

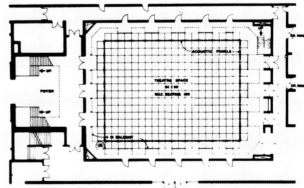

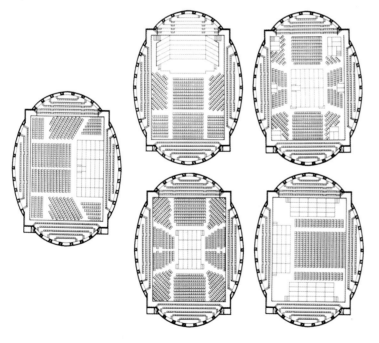

means of a broad, skylit ramp and stairway (worthy of the Palais Garnier) that spirals gracefully around the elliptical concert hall to an entrance at the far side.

The flat orchestra level has a low balcony on three sides and boxes at the top periphery. Blocks of two, four, or five rows of three seats each sit on pallets supported by air cushions that operate on the same principle as a hovercraft: once the cushions are activated, a couple of stagehands can move the seating into various positions. The theater undergoes twenty to twenty-five reconfigurations annually, despite union restrictions that require more stagehands than necessary to transform the space in a day. Laurent Bayle, general director of the Cité, acknowledges the difficulties of spaces designed for experimental music: "Today composers want spatial options, but they are inhibited by their fear that there is not enough public for such works."[73]

Portzamparc's postmodern architectural style is arguably overly aggressive, but what matters is that, despite forewarnings of an acoustical disaster, in the hall's original form music sounded good there. (Albert Yaying Xu was the acoustician.) Mark Swed, the chief classical music critic of the *Los Angeles Times*, has said of the auditorium: "I heard a concert with various world musicians some years ago and found the space to be immediate, [with] no sense of distance between stage and ears."[74] (The addition of numerous lighting grids around 2001 added sound absorption, thereby inhibiting the lengthening of reverberation time, although the reverberation can be reduced for contemporary and world music.)[75] The Cité de la Musique auditorium was deemed so satisfactory that it was taken as a model for a similar modular theater, the Judy and Arthur Zankel Hall (2003) by Polshek Partnership (now Ennead Architects) at Carnegie Hall (1891) in Manhattan.

Zankel Hall, in whose planning Boulez also participated, was the brainchild of the late Judith Aaron, former executive director of Carnegie Hall, and Isaac Stern, who saved the historic hall from demolition. The hall replaced a basement theater that had been part of the historic building. It had served rarely because of a lack of acoustical isolation between the lower and upper auditoriums. The space, eventually occupied by the Carnegie Cinema, was enlarged at tremendous expense by excavating the bedrock below the main Isaac Stern Auditorium.

There is a big difference between the ways the Cité and Zankel are used. Stagehands in New York earn at least twice as much per day as comparable employees in Paris, and cultural institutions in the United States are without the hefty government subsidies that have been routinely enjoyed by those in France. As a result, New York concertgoers would be hard put to know that changes can be made in the disposition of the theater's seats and stage.

Zankel's dreary Seventh Avenue entranceway, with escalators descending two levels, does not do justice to the handsome, modernist, below-grade hall. An elliptical concrete enclosure whose polished plaster walls define the lobbies contains a

Cité de la Musique, Paris, France, 1995. Exterior view; concert hall; plans; lobby.

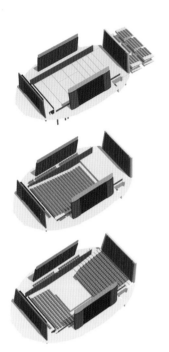

rectangular auditorium with hospitable honey-colored American sycamore paneling and bronze detailing. The auditorium has up to 644 seats positioned in groups of twelve on chair wagons; these move on the same kind of air cushion as those at the Cité de la Musique. Additionally, nine lifts allow modification of the floor. Interactive videoconferencing projected on the hall's screen is technology that has been adopted in a number of venues.

In spite of Zankel's ability to shift from end to center stage or flat floor in a day, with rare exceptions it has remained in the same proscenium mode since its inauguration, even when, under Guzelimian's watch, *Répons* was performed there. In explaining why this theater's potential flexibility has not been implemented, Carnegie Hall's executive and artistic director Clive Gillinson says, "Zankel is not about architecture; the architecture serves the music."[76] Gillinson's response is puzzling, because changes in the hall—as Venice's San Marco changed to accommodate the new antiphonal music in the sixteenth century—would in fact make the architecture serve the music.

Even more of a question is the fate of the modular space that was envisioned for the Bastille Opera in Paris (1989). As in the case of the Cité de la Musique, the

1.5-million-square-foot, $570 million opera belongs to the government's series of *grands projets*; the modular venue was part of the scheme that won Carlos Ott the open architectural competition to design the new building. Planned for an audience of approximately 1,500, this experimental theater would have offered a welcome contrast to the impersonal 2,700-seat main auditorium and sounded a note of innovation within the otherwise pedestrian main opera house. Boulez figured among a group of progressive cultural professionals, including Daniel Barenboim, then chief conductor of the Bastille Opera, and Gerard Mortier, the incoming general director, who were eager for such a facility.

It was not to be. Although remnants of the medium-size theater are still visible, few of those arriving at one of the main portals in the Bastille Opera's already worn granite, stainless-steel, and glass cylindrical facade notice the permanently closed entranceway to the south of the main one. It leads to the small lobby of the projected modular theater, which adjoined, and was slated to share, the backstage facilities of the main auditorium.

Taking advantage of the small theater's incompletion when the big house was inaugurated, Hugues Gall, then general director of the opera, backed by the political appointee Pierre Bergé, cofounder of the Yves St. Laurent fashion house, rejected the project in favor of a rehearsal space. Consequently, the Salle Rolf Liebermann sits sadly within an enormous raw space in which an unfinished balcony and fly tower bear ghostly witness to the aborted project. If ever there was a phantom of the opera, this is it.

Bastille Opera House, Paris, France, 1989. Remains of modular theater; axonometric showing modular theater to the right of main auditorium.

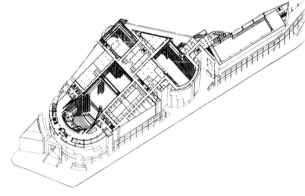

TO THE HUGUES GALLS AND CLIVE GILLINSONS OF THE MUSIC WORLD, WHO
defend the status quo, I would ask if modular theaters and spatial options can be considered far-fetched when already in the sixteenth century Adrian Willaert and the
Gabrielis had distanced from one another two organs and two choirs at San Marco.
Even though the resumption of ideas from the past and new inventions came slowly,
they have occurred with increasing frequency since the 1960s, and they are accelerat-
ing in the first decade of the twenty-first century.

Certainly, there is room for many options. James Conlon, who among his diverse
duties serves as music director of the Los Angeles Opera, asserts "a screaming need for
smaller opera houses, suitable for Baroque, classical, and those contemporary operas
conceived for intimate spaces."[77] But at the same time, he concedes, "No sane individual would recommend destroying the Musikverein, Bayreuth, or La Scala." Those
historic theaters were at the time of their construction major expressions of social and
cultural values—as theaters have been since antiquity, and remain to this day.

The history of theater design, and of the opera houses and concert halls derived
from theater, reveals a variety of inventions that connect spectators and performers.
A long and venerable lineage extends from Palladio to Wagner, the renowned stage
designer Adolphe Appia, Boulez, and Mortier, all of whom looked to the theater
of ancient Greece as a model. For his own part, Mortier asks, "Why not do away with
elaborate, outmoded sets and allow the architecture to speak for itself? One prop
can be enough."[78]

Musicians are seeking ways to intensify their relationship with the audience.
Dramatic lighting and sophisticated video projections are increasingly important for
concerts as well as opera, as are off-stage areas for performers. More and more, these
and other theatrical effects accompany presentations of new music (which frequently
incorporates such devices in the scores) and pieces from the classical repertoire.

"I love singing in halls with a history," says the great American soprano Renée
Fleming, "but new opera reflects our time, so the space should reflect this."[79] Robert
Lepage's staging of *Das Rheingold*, premiered in September 2010 at the Metropolitan
Opera in New York City, exemplifies the beginnings of digital capabilities. In the
opera's first scene, digital technology made the bubbles that seemed to stream from
the Rhinemaidens' mouths increase in intensity according to the loudness of their
singing. Eighth Blackbird is one of several contemporary classical music groups
whose members engage in choreographed movement around the stage while giving
a concert in order to highlight musical affinities and to enliven the show.

Unquestionably, Wagner's Festspielhaus offered a more democratic seating
arrangement and encouraged a more immersive listening experience than opera
houses had up until then. But for all of its revolutionary characteristics, the Festival
Hall retained, and even emphasized, the separation between actors and audience

that is criticized today by many. The theater director Peter Sellars expresses a common goal: "Instead of a theater of isolation in which we bury ourselves alive, it is now about engagement, communal participation, and exchange in shared, vibrant, alive spaces."[80] Yet Sellars acknowledges the complexity of these issues. In contrast to his call for audience participation in certain events, for opera Sellars asserts, "I can't stand theater in the round because I don't want to see backs, and eye-to-eye connection with the stage is crucial."

The innovations of the twentieth and twenty-first centuries—important as they are—nonetheless remain rare exceptions to the traditional architecture that has been built for music since the 1950s. Reconstruction in Europe after World War II favored conservative architecture, and in the United States spaces like the Henry and Edsel Ford Auditorium in Detroit (1958) and the enormous Arie Crown Theatre in Chicago (1960) typified the blandness of that era. It was then, too, that Lincoln Center for the Performing Arts in Manhattan launched the construction of equally conservative cultural centers across the country.

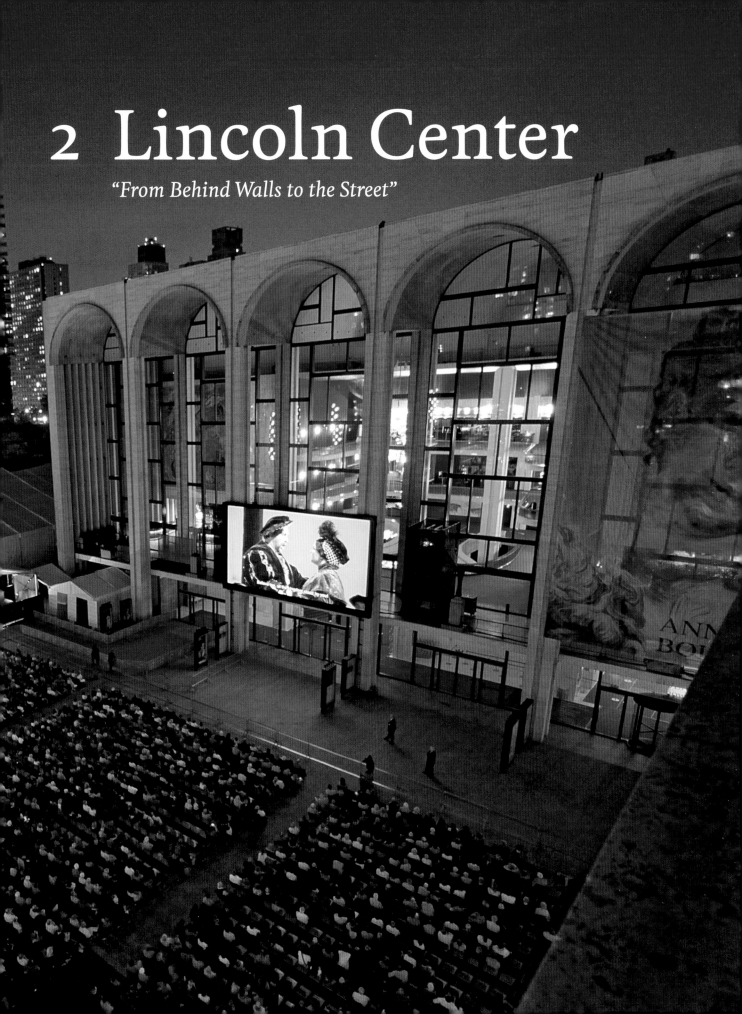

2 Lincoln Center

"From Behind Walls to the Street"

T

HE BUILDINGS THAT TOGETHER MAKE UP LINCOLN CENTER on Manhattan's Upper West Side were designed and constructed between 1957 and 1969. The principal architects of the twelve constituent organizations were some of the most influential figures of their generation—Wallace K. Harrison (Metropolitan Opera), Max Abramovitz (Avery Fisher Hall), Philip Johnson (David H. Koch Theater), Eero Saarinen (Vivian Beaumont Theater), Gordon Bunshaft (Library for the Performing Arts), and Pietro Belluschi (Juilliard School and Alice Tully Hall). Despite the participation of these prominent personalities—and in part because of their competitiveness with each other—the Lincoln Center theaters have been the butt of negative criticism from the moment they were completed. Their architecture and the acoustics of their auditoriums (except for the Metropolitan Opera House and Alice Tully Hall, to a point) were deemed qualitatively inferior, and urbanistically the center has failed the test of time.

Throughout this chapter, individual theaters are referred to by their current names.

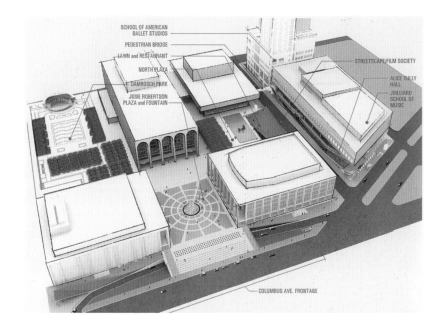

Lincoln Center's architects thought of it, variously, as "an area isolated from the hubbub of New York City," "a fortress," and "a casbah with high walls to the outside world."[1] It was a concept perfectly fitted at the time to the sixteen-acre superblock made available for development by New York City Parks Commissioner Robert Moses, yet already retrograde from a planning perspective. By the 1950s forward-looking urbanists like Jane Jacobs were advocating more contextual models.

Nevertheless, such superblocks were a favorite leitmotif of 1960s urban renewal, even though they have no through streets or active ground-level frontage, and their isolation is questionable. Placed within a superblock, elevated like temples on a plinth, Lincoln Center's buildings were removed from the life of the city around them and insulated from the low-income housing behind, on Amsterdam Avenue. A similar attitude informed the selection of a riverside site near the edge of Georgetown in Washington, D.C., for the John F. Kennedy Center for the Performing Arts, which is inhospitable to pedestrians and is serviced by a single underused bus line.[2]

This idea of exclusivity applied to art, which still held sway in the mid-twentieth century, was by the turn of the twenty-first century incompatible with efforts to make all the arts part of everyday living. Early in 1999, the Lincoln Center constituents began to plan a series of renovations. With the help of Beyer Blinder Belle, an architectural firm known for its preservation work, the group drew up a list of repairs, improvements, and special projects; the estimated cost was $1.5 billion.[3] Alice Tully Hall, the David H. Koch Theater, and Avery Fisher Hall were to be renovated (along with the Vivian Beaumont Theater, which is not used for classical music). As important as the work to be done on the buildings were the enlivening changes and

Lincoln Center for the Performing Arts, New York, New York. Aerial perspective, 2010; aerial view, 1969.

additions slated for West Sixty-fifth Street and Columbus Avenue, intended to make Lincoln Center's relationship with them more inviting and more inclusive.

The renovations that have been completed and those still under consideration encapsulate some of the profound shifts that have taken place in attitudes toward the city, toward the arts, and toward the process of theater design since the center was founded. The changes are not simple matters of differences in architectural taste but embody questions of cultural and social transformation that affect urban placement, the complexities of user needs—for audiences and performers alike—the contentious, paramount subject of acoustics, and the way in which technical advisors interact with clients and architects.

Official reasons for the massive 1999 reevaluation included "the need for repair and renovation."[4] The buildings had, after all, been functioning for over three decades. But as vital as correcting the physical deterioration and accounting for the technological progress that could be expected for that amount of time was a radical shift in urban thinking.

Great historic European opera houses—think of the Palais Garnier in Paris, La Scala in Milan—are part and parcel of the urban fabric (see Chapter 1). And in the last quarter of the twentieth century, new opera houses and concert halls, together with other cultural institutions, began to connect even more directly with the public, shedding forbidding formality and opening to the exterior. The 1977 Centre Culturel Georges Pompidou in Paris (Renzo Piano and Richard Rogers) led the movement for museums; Jean Nouvel's 2000 glass-sheathed KKL in Lucerne represents the trend in concert halls.

As considered from the more open-minded, inclusive zeitgeist of the western world in the new millennium, even the Manhattan neighborhood's initial blighted condition didn't justify Lincoln Center's exclusiveness. Today, cultural facilities are designed to reach out to depressed surrounding areas, as the Pompidou did in the 1970s and as the Walt Disney Concert Hall (Frank Gehry, 2003) does in what had become the intimidating corporate downtown of Los Angeles (see Chapter 3). Now that Lincoln Center's surroundings have been transformed by fifty high-rise buildings, upscale commerce, and residential uses as well as other entertainment offerings, there is even less excuse for its isolation.

The goals and tastes of the civic-minded people who realized Lincoln Center were more in tune with the conservative viewpoints that governed New York City's late-nineteenth-century Metropolitan Opera House project than with the twenty-first century's more liberal attitudes. The original Metropolitan Opera House (designed by Josiah Cleveland Cady), which opened in 1883 at Broadway and Thirty-ninth Street, had been faulted since day one on every count, including poor sight lines, inadequate storage space for scenery, and a firetrap balcony. Despite the building's redesign by

Carrère and Hastings in 1902, by 1918 a new opera house was seriously under consideration, and four years later the wealthy businessman Philip Berolzheimer launched the idea of combining it with other theaters in a center for the performing arts—the kernel of the idea for Lincoln Center. Over time, various attempts to realize a cultural center in a number of mid-Manhattan locations all came to naught.[5]

What finally made the center possible was the determination of the all-powerful city parks commissioner (and head of the State Parks Council) Robert Moses to make available real estate on Manhattan's West Side that answered both the Metropolitan Opera's need to rebuild and the New York Philharmonic's threatened eviction from Carnegie Hall by a projected office building. It was the leaders of the opera—Charles M. Spofford, chairman of its executive committee, and Anthony A. Bliss, its executive director, plus their counterparts at the New York Philharmonic (then the New York Philharmonic-Symphony Society), Floyd Blair and Arthur A. Houghton, Jr.—who were the key players, together with John D. Rockefeller III, whose $500,000 donation helped to spearhead the enormous fundraising drive that he eventually led. This initial group of patrician white males was soon joined by leaders of other institutions who had similar backgrounds.

Many could be described as Herman Krawitz, the opera's efficient business and technical manager, described Bliss: he was a successful lawyer who enjoyed wide-ranging experience with cultural institutions, but "he was more social."[6] Edgar B. Young, who worked closely with Rockefeller, called the Lincoln Center directors "the elite establishment at work . . . at its best [in intent] . . . or, a critic might say, at its worst [in their remoteness]."[7]

The appointment in 2009 of Katherine G. Farley as chair designate and her subsequent chairwomanship of Lincoln Center illustrate changes in the makeup of the boards, from a plutocracy to a quasi-meritocracy. (Her appointment was preceded by Joseph Volpe's designation as general manager of the Met; he was the first to rise to this position from within the ranks of the company.) Bliss's involvement with the center continued his father Cornelius's connection with the opera, on whose board the senior Bliss served from 1932 until his death in 1949. In contrast, Farley came to the board as a successful businesswoman, now senior managing director of Tishman Speyer, where she has been since 1984. Farley's professional affiliation with New York City's largest real estate company underlines the project's setup as a real estate operation, with the constituents renting theaters from Lincoln Center for the Performing Arts, Inc.[8] That Farley's husband, Jerry I. Speyer, happens to be a chairman and co-chief executive of Tishman Speyer cannot be overlooked.

Though traditional in their tastes, the center's founders were cultural pioneers. Few large concert halls were built anywhere in the world in the first half of the twentieth century, and European efforts to repair the devastations of that war had produced

opera houses that were criticized by forward-looking music professionals for appearing very modern from the outside while remaining extremely old-fashioned on the inside.[9] Not only did the New York philanthropists build the first such grand, inclusive aggregate, but within a few years Lincoln Center had inspired similar complexes across the United States, taking this country to a new level of cultural sophistication. For this they deserve the highest praise. But at the same time, many of the ways in which they approached the project were misguided. In the absence of official constituents for the Koch and the Vivian Beaumont theaters, there was no clear representation of user needs for these buildings. A labyrinth of committees and subcommittees pitted the various organizations' ideals against Lincoln Center's efforts to control costs and compromised the clients' relationship with the architects.[10] Unrealistic estimates necessitated endless cost cutting that drained the initial schemes of their energy, as in the case of Harrison's reinterpretation of Bernini's colonnade at St. Peter's in Rome. (Johnson, however, managed to carry out his Roman Campidoglio motif for the central Josie Robertson fountain plaza.) In the founders' defense, theater design in the 1960s did not benefit from the expertise that is available today, including that of acousticians and professional theater consultants, to which the center's creation made a valuable contribution.

The establishment network in charge chose a comparably establishment architect to design the opera house. And since the Metropolitan Opera was the most powerful of the participating institutions, that architect, Wallace K. Harrison, also headed the project. Harrison had been associated professionally (and through marriage) with the Rockefeller family: his office was one of the three architectural firms responsible for Rockefeller Center (1933). Later, as a principal of his own firm, his conciliatory and fair-minded nature served him well as the head of the team effort responsible for designing and constructing the United Nations Headquarters (1952) beside the East River. Faced, however, with a similar group of assertive architects at Lincoln Center, the mild-mannered Harrison balked at assuming the same stressful role of general arbiter.

Even before the architects had been designated, an advisory group selected from Harrison's suggestions, which included modern masters such as Alvar Aalto and Marcel Breuer (who were later dropped), bitterly criticized Moses's determination to reserve the area's southwest corner for yet another of the many parks he created throughout New York State (a criticism that is justified, it would seem, by the now-deserted condition of both the park and West Sixty-third Street). The consequent reduction of the site was regarded by the architects as detrimental to the plan, and their acrimonious objection to it set the tone for future meetings. Compounding discord among the architects were the rigorous restrictions imposed by Lincoln Center for the Performing Arts, Inc., such as the requirement that the twelve constituents unanimously approve every public space change or change affecting the campus. As the enterprise progressed, Harrison left more and more decisions to his colleagues and to the committees, with devastating results. Key committee members—Spofford and Bliss at the Met, for example—in turn deferred to Rockefeller, who proved to be as indecisive as Harrison.[11]

As for the six architects who were selected to work at the complex, only Belluschi and Saarinen had proven themselves innovative at this level of practice. Harrison, Abramovitz, and Skidmore, Owings & Merrill's Bunshaft were known for their expert handling of large-scale commercial architecture rather than for creative flights of fancy. Johnson was a maverick; a reputed tastemaker, his work was often simply trendy, and in the late 1950s he had already begun to flirt with one of his least convincing styles, what later became known as postmodernism, which informs the whole scheme. Altogether, the design by committee to which the architects were subjected at Lincoln Center produced, in the words of Ada Louise Huxtable, the dean of architecture critics, "soft modernism."[12]

West Sixty-fifth Street and the Juilliard School

Pietro Belluschi in association
with Eduardo Catalano and Helge Westermann, 1969
Renovation, Diller Scofidio + Renfro, 2009

The Lincoln Center Development Project, Inc., was in place by the end of 1999. Under the direction of Rebecca Robertson—who as president of the 42nd Street Development Project, Inc., had skillfully managed the regeneration of Times Square in the 1990s—emphasis was given to the buildings' relationships with the streets they face. First to be addressed were the areas and buildings on both sides of West Sixty-fifth Street—the Juilliard School with Alice Tully Hall at the north, the streetscape at the south—and access to the central fountain plaza at Columbus Avenue. Diller + Scofidio (which became Diller Scofidio + Renfro in 2004) was entrusted with the renovation in collaboration with FXFOWLE Architects; Cooper, Robertson & Partners for planning; L'Observatoire International for lighting; Olin Partnership for landscape architecture; and 2x4 for graphic design and visual identity.

Just as the character of the current Lincoln Center directors differs from that of their earliest predecessors, so too is the work of the two founding principals, Elizabeth Diller and Ricardo Scofidio, worlds apart from that of the initial architects.[13] Their unrealized Slow House (1991), a Long Island retreat, electronically transferred a magnificent ocean view to an interior monitor in the kind of technological sleight of hand favored by these architects. In 2003, when Diller + Scofidio received the cultural center commission, the firm had had more attention for its Blur Building of the previous year—a temporary exhibition pavilion on Lake Neuchâtel in Switzerland consisting primarily of self-generating fog—than for its single completed project, social housing in Gifu, Japan (2000). With so few built works to its name, the office had for many years based its reputation on theoretical exhibitions, installations, and performances. The latter in particular are still a strong influence on DS+R's architecture, which displays a keen sense of how bodies move through space and how people inhabit space visually.

This instinct for movement informs Diller Scofidio + Renfro's alterations to the main entries to the center. The inhibiting drop-off road paralleling the daunting hurdle of Columbus Avenue and Broadway's six traffic lanes (and built to support the car-centric culture of the 1960s) has been sunk to the concourse level and covered over with a ceremonial entrance to the central fountain plaza. This wide stairway has been made more attractive (and more difficult to navigate for some) by deeper steps and lower treads. Animating the stairway and lighting it at night are announcements

Columbus Avenue
entrance, renovation,
2010.

of upcoming events spelled out like text messages by LED lights behind the risers' etched glass (the Sixty-fifth Street stairway features the same treatment). Less success-ful are the knifelike glass canopies that extend to the sidewalk from the arcades of Avery Fisher Hall and the Koch Theater. While the ramps and railings they overhang are helpful to many, the extremely narrow width of these porte-cocheres impairs their sheltering function and appears skimpy in relation to the plaza.

Facing West Sixty-fifth Street and Broadway, the Juilliard School was the third major institution to join the center, but it was completed last, since it had taken fifteen years to reconcile the building's budget and its exceptionally complex pro-gram (including four performance halls, classrooms, rehearsal rooms, and offices). Juilliard's primary architect, Pietro Belluschi, had a fine reputation for innovative modern architecture—one example is his Equitable Building in Portland, Oregon (1948), an early and particularly successful glass-and-aluminum skyscraper. He was, however, repeatedly asked to revise his design for Juilliard as plans to develop the block north of the campus as another superblock were gradually abandoned by the board of directors.

Along both Sixty-fifth Street and Broadway, the extensive, austere facades of Juilliard (a steel-concrete composite system in a fairly regular grid) were clad in travertine to match those of all the Lincoln Center architecture. Linking the build-ing with the center was Milstein Plaza, a broad pedestrian overpass above Sixty-fifth Street. It was severely criticized by Bunshaft and Saarinen, both of whom presented alternative schemes for narrower walkways that were overruled by Abramovitz, who was in charge of supervising Belluschi's design. Abramovitz backed the Milstein

Elinor Bunin Munroe Film
Center, 2011.

Peter Jay Sharp Theater
and Alice Tully Hall,
renovation of Sixty-fifth
Street facades, 2010.

bridge because it buffered the center's buildings from the sight and sound of traffic on the street below.

Sixty-fifth Street was an initial focus of DS+R's attention, and Belluschi's hefty pedestrian plaza, which turned the street into a traffic tunnel, was the first thing to go. It is a relief to once again be able to enjoy daylight and openness on the center's spine. In the spirit of Saarinen's project for two slender, bronze-covered bridges, DS+R's Malkin footbridge (2011) is narrow (ten feet wide in comparison to the Milstein's two hundred feet). Partially in steel, the walkway meets the main campus diagonally, as does a structural support.

To the south, the black hole–like entrance to the center's parking garage, the garage itself, and some offices have been replaced by the Elinor Bunin Munroe Film Center. The architect and designer David Rockwell, known for his theatrical settings, worked with DS+R to produce the bright orange entryway, with sparkling LED lights underfoot, and a fritted-glass facade etched with the titles of a thousand films that have been shown by the Film Society of Lincoln Center. LED announcements run above the orange "marquee," in dialogue with the changing electronic billboards that dot the pavement: texts, color, light, and movement animate the thoroughfare. Within, Rockwell executed two screening rooms—with 90 and 150 seats—a café, and a lobby that can be used as an 87-seat amphitheater. He says that "giving the Film

Society this public face makes it part of the community."[14] The roadway has been restricted from three to two lanes in order to widen the sidewalk on the south side of Sixty-fifth Street. A double staircase from the sidewalk to the North Plaza invites visitors up. On the north side, masonry facades have been replaced with glazing, opening the lobbies of the Peter Jay Sharp Theater and Alice Tully Hall to the street and revealing a new, friendlier entrance lobby to the Juilliard.

Diller Scofidio + Renfro's 220-seat restaurant, Lincoln, in the North Plaza is another instance of the architects revisiting the first design process. Warning at the time that the campus lacked urban vitality, members of the building committee repeated the criticism voiced by the respected architectural critic Frederick Gutheim, who said that the center would be isolated from the amenities of urban life; to avoid this, they encouraged the inclusion of shops and eating establishments along Juilliard's Broadway frontage. Such commercial enterprises were roundly vetoed by the center's architects, backed by Peter Mennin, the young composer who became president of the Juilliard in 1962, on the grounds that they would compromise the purity of purpose of a cultural facility—and the purity of its architecture.

What was considered pure fifty years ago, though, is today considered arid. Rather than aspiring to transcend mundane daily routines, many art forms try, rather, to be part of them. The great majority of those involved with promotion and publicity perceive culture as an activity that includes the enjoyment of food and of social amenities, all of which help to attract the public. Placed behind Avery Fisher Hall and fronting Sixty-fifth Street, the Lincoln restaurant's playful parabolic, grass-covered roof—the Laurie M. Tisch Illumination Lawn—beckons to those on the North Plaza and is already a favorite place for sun worshipers. (It does, however, crowd the plaza's Henry Moore sculpture and reflecting pool, destroying the landscape architect Dan

Malkin footbridge, rendering, 2011.

Lincoln restaurant, 2010.

Kiley's handsome setting for the Vivian Beaumont Theater.) The steel-beam and layered-concrete dining facility protects the plaza from street noise, as the Milstein Bridge did. The roof extends beyond the restaurant's glass walls, making it visible from Broadway, and the floor-to-ceiling glazing gives diners great views to Broadway and to the plaza, while revealing the interior (including the kitchen) to passersby on the street.

When the Juilliard School and the Beaumont Theater were completed, Huxtable declared them "better buildings than the tiresomely, tentatively traditionalized star structures."[15] Granted, Belluschi at the Juilliard and Saarinen at the Beaumont had resisted the tedious neoclassic modernism that was adopted for the Met, Avery Fisher Hall, and the Koch Theater. Today, however, Diller calls the Juilliard's style "brutalism in drag,"[16] and indeed for pedestrians used to the school's stolid demeanor, the building's new transparency is a welcome change. When Huxtable visited the first building tackled by DS+R, she pronounced the Juilliard School's new glazed Broadway facade "a miracle from the street."[17] This is especially true of its southeast corner where, framed by a dramatic cantilever toward Sixty-fifth Street, the nearly forty-foot-high lobby of Alice Tully Hall is visible. Only the small outdoor grandstand at the uppermost corner of the sunken plaza surrounding the theater strikes a discordant note, appearing out of scale with the massive building it faces. To be effective, it needs

Juilliard School, 1969. Exterior view; renovation of Broadway facade with Alice Tully entrance (at left), 2010.

to be filled with people as it was in the summer of 2009 for a performance by Asphalt Orchestra, a high-art street band.

Diller says the architects "wanted to bring the Juilliard's activities from behind walls to the street," and they have certainly succeeded. The cantilevered $7.2 million Lincoln Kirstein Studio wing added to the building, whose underside frames the Tully lobby, is clad entirely in glass and appears particularly festive at night when it is illuminated. Hanging under the addition's prow is a wide, glass-fronted box in which, when the shades are up, passersby can glimpse the Juilliard dancers at work, and they in turn can see the street. The concept recalls the box that DS+R suspended below a similar giant cantilever at the Institute of Contemporary Art in Boston (2006), where visitors in the museum's raked mediathèque look out to the seascape.

The Juilliard's approximately 45,000 square feet of gained space includes a large rehearsal room and six teaching studios built into what were formerly two court-yards. DS+R also remodeled a portion of the Samuel B. and David Rose Building (Davis Brody & Associates, now Davis Brody Bond Aedas, 1990). The architects gained 1,200 square feet by nesting two dance studios in the existing space (the old ceilings were dropped from 16 to 11½ feet), which was heightened by rerouting to the periphery ductwork that had occupied a deep ceiling plenum. Both new studios have uninterrupted spans and floor-to-ceiling glass; the inserted upper studios appear to float in space due to walls that do not touch those of the perimeter. Between the old and new studios is a mezzanine lounge with liquid crystal walls that can be switched electrically from translucent to transparent glass. The design opens the former envelope of the Juilliard to the street and from one interior space to another, as the ingenious use of electronic signage does conceptually as well.

Samuel B. and David Rose Building, 1990. Renovation of dance studios, 2009.

Institute of Contemporary Art, Boston, Massachusetts, 2006.

Alice Tully Hall

Within the Juilliard School

Alice Tully (1902–93) was the daughter of a New York State senator and the grand-daughter of Amory Houghton, the founder of Corning Glass Works, whose fortune she inherited. She made her debut as a professional singer in 1927 in Paris and continued her career on both sides of the Atlantic until about 1950. Tully's cousin Arthur Houghton, Jr., persuaded her to become involved with Lincoln Center, and her generous gifts to both the auditorium ($4.5 million) and the Chamber Music Society entitled her to the last word for the hall, from the decor to the acoustics.

To achieve the building's new transparency, Diller Scofidio + Renfro stripped away the wide staircase to the bleak, windswept Milstein Plaza, which detracted from the theater's entrance, and replaced the masonry walls on Broadway and the bottom part of the Sixty-fifth Street facades with mullionless glass, thus connecting the interiors with the street. The original Alice Tully lobby has been compared to an air-raid shelter,[18] and even Belluschi had been unhappy with the bush-hammered concrete finish of the walls that he was forced to use instead of wood, for economy's sake. Now almost doubled in size, the lobby has not only become a festive, light-filled public space, but for the first time it clearly signals the presence of a theater. Café tables are scattered attractively in front of a sleek limestone bar curving along the brownish-red back wall of Brazilian muirapiranga wood that continues into the theater lobby. One serious downside to the stunning renovated space is the box offices—they are squeezed into a small northeast corner too far from the hall. And with the café full, excessive noise is also problematic.

Within this glamorous container, the auditorium itself, now called the Starr Theater, also looks dazzlingly different from its earlier incarnation. Limited as DS+R was to the existing structure, the $157 million renovation could not change the theater's shape. It did, however, transform its look, upgrade the acoustics, and make the stage configuration more flexible.

The original hall's vertical wood sound diffusers have been replaced by red-brown African moabi in wide segments that step back from the stage. This thin, fine-grained wood veneer is heat-bonded to resin panels on the stage walls and side walls of the auditorium; behind large expanses of the translucently thin moabi, tiny LED lights embedded in the panels create a glow that can be adjusted from pale orange to deep red.

What Diller calls the room's "blush" recalls a similar concept used by the German architect Heinrich Tessenow, in collaboration with the stage designer Adolphe Appia, for one of the first great modern theaters at the Institut Jaques-Dalcroze, an ideal

planned community at Hellerau, Germany (1911). Without today's sophisticated technology, Tessenow placed thousands of electric light bulbs between two layers of white
canvas impregnated with wax, covering the walls and creating a variable diffused
light in different hues.[19] Appia simplified the stage design to emphasize the actors'
movements, which he coordinated with light and music. The continuous space of this
early-modernist landmark was an example of early-twentieth-century attempts to
bridge the separation between stage and public. At the Starr Theater the architects did
their best to achieve something comparable by using the "blush" to unify the stage
and auditorium walls, making the theater more intimate for both sight and sound.

Intimacy depends on many factors. Closeness of seats to the stage comes into
play, as do a number of intangibles. Diller uses the term "intimacy" to designate protection from foreign intrusions on the theater experience. The auditorium is now a
sound cocoon. It is insulated from exterior interference, like traffic and the subway, by
a partial box-in-box construction with isolating walls and slab; interior noise caused
by audience footsteps and conversations and also the mechanical systems is reduced

Lobby renovation, 2009.

by a wood liner of the walls, floor, and stage. Unwanted elements such as railings have been removed so that, as Diller points out, "there is no visual noise." A new system of automated modifiable extensions that allows three projections of the stage into the hall (instead of one fixed stage size) also contributes to greater intimacy, as does the redesign of the seating areas.

A traditional balcony running straight across the room, with two small rectangular balconies stepping down from it at either side, has been replaced by a monolithic seating area. A curved balustrade swoops down toward the stage, its shape, like every design decision for the auditorium, guided by acoustical considerations. But because it was too difficult to remove the original side balconies, the new balcony simply incorporates them: clunky concrete boxes that poke above the streamlined form. The balcony's central aisle and the section divider aisles that ran across the parterre have been eliminated, increasing seating capacity from 923 to 1,087 and leaving uninterrupted continental seating in both areas. Thanks to Alice Tully, a tall woman who insisted on ample legroom, the rows are spaced far enough apart to allow relatively easy access.

Lavender carpeting (and raspberry upholstery) chosen by Tully has been replaced with moabi wood in the parterre aisles and concrete in the seating areas (now with gray upholstery) and throughout the balcony. The industrial carpeting and felt-covered walls of the passageways leading to the auditorium's side entrances are gray. These cheaper materials do not detract from the overall effect (nor do any of the other cost savers, like somewhat imperfect exterior detailing), although better lighting in the passageways and near the balcony steps would facilitate circulation.

The press greeted Alice Tully Hall's reopening in February 2009 with praise for the renovated entranceway, refurbished auditorium, and acoustics. Anthony Tommasini, the *New York Times* chief music critic, titled his review "At Last Heavenly Acoustics Are Heard in the Hall."[20] Barrymore Laurence Scherer in the *Wall Street*

Institut Jaques-Dalcroze, stage, Hellerau, Germany, 1911.

Juilliard School Dedication Concert, Alice Tully Hall, October 26, 1969.

Starr Theater with "blush," 2009.

Journal declared that "the sound of Tully Hall is now almost tangible . . . you feel the sound as if it were a three-dimensional entity."[21] But both critics prefaced their endorsements of the altered acoustics by saying that they had never had a problem with the old. Indeed, responding to the hall's inauguration in 1969, Alan Rich, the music critic for *New York Magazine*, described the hall's sound as "warm, rich and clear,"[22] and Harold C. Schonberg, at that time chief music critic of the *New York Times*, wrote, "I have not heard a dissenting voice about Tully Hall. Everybody seems to like its intimacy, smooth blend of sound and reverberation characteristics."[23]

Given this general approval of the hall at its inception, why, over the years, was there increasing criticism of its acoustics? And why, given the initial enthusiasm for the renovation, did Allan Kozinn of the *New York Times*, within a few months of the

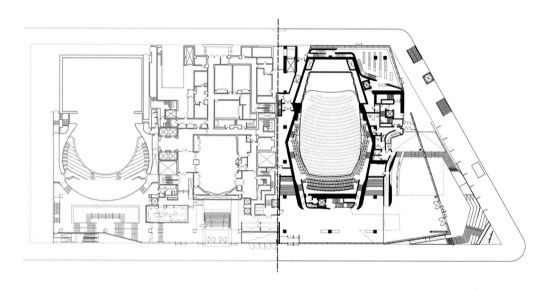

Juilliard School, plan with Peter Jay Sharp Theater and Morse Hall (left and center; not part of the current renovation) and Starr Theater, 2009.

Starr Theater's inauguration, attack the refurbishing and in particular the acoustics, which he described as "arid, with no resonance at all"?[24] (Kozinn appears to have taken up, against Alice Tully Hall, the cudgel that was wielded relentlessly in the 1960s and 1970s by Schonberg against Avery Fisher Hall.)

J. Christopher Jaffe, of JaffeHolden (and since 2009 senior consultant for Acentech's Studio A in Boston), the firm that served as acoustical consultant for the renovation, offers one explanation. The relatively dry reverberation of the hall as it was first built, asserts Jaffe, was well suited to smaller ensembles, especially when they were positioned on the stage's forward lifts. The Chamber Music Society of Lincoln Center is Tully's prime tenant. When an orchestra was booked into the space with the musicians playing symphonic music, the environment was less satisfactory.[25] It was in fact the large Deutsche Kammerphilharmonie Bremen's performance of Beethoven that Kozinn singled out as revealing "unmistakably the flaws in Tully's sound," although he insists the problem exists for all types of performance in all three stage configurations.

Jaffe's explanation is backed up by at least one musician who has performed in the hall. On the basis of five concerts in the renovated hall—two for the Mostly Mozart Festival and three for a Dvorak series—Eugene Drucker, a violinist with the renowned Emerson String Quartet, says, "There is no question that there is a significant improvement, more presence and core to the sound; before the renovation, unless the repertoire was very forceful (Bartok or Shostakovich as opposed to Haydn), the performance didn't seem to have enough impact. Both the visual and the audio aspects are now more engaging."[26]

For particularly loud, bright music, the acousticians provided sound-absorbent banners that can be dropped along the walls to dampen reflections of the sound, to avoid slapback (doubling the attacks of sounds) from the walls, and/or to cut brightness. Banners were used for an all-Xenakis program (March 5, 2011) in which sustained loud passages remained clear and did not overwhelm the room. Concert halls are often compared to a musical instrument with which the player must become familiar in order to maximize the effect. It remains to be seen how the Starr's musicians will learn to play this hall.

Additionally, acoustical perceptions change over time. A good example of this is provided by the late Heinrich Keilholz, acoustical advisor for Alice Tully Hall, who had performed the same function for Vienna's Staatsoper when it was rebuilt after World War II (it reopened in 1955). When Keilholz and his team in Vienna were asked whether they could duplicate the excellent acoustics of the 1869 opera house, they questioned the desirability of such a goal: high-fidelity recordings had changed the way people listened, and he was convinced that what had been satisfactory before the opera's destruction in the 1940s would no longer be so in the mid-1950s. (A similar change in listening expectations took place in the last third of the twentieth century, when meticulously edited digital recordings offered a degree of "near-perfection" that is unattainable in a live concert.) In the 1950s, Keilholz explained, "Elderly gentlemen who play chamber music at home . . . don't like our new, bright-sounding halls. But the truth is that many chamber groups are coming to prefer medium-size, bright-sounding halls . . . It's all pretty much a matter of what one is accustomed to."[27] Shifts over time in the appreciation of sound at Alice Tully Hall, as well as at other theaters at Lincoln Center, seem to confirm Keilholz's statement about this highly subjective issue.

David H. Koch Theater

Formerly New York State Theater
Philip Johnson, 1964
Renovation (most recent), JCJ Architecture, 2009

The late David Taylor, a theater consultant with Arup Performing Arts, insists that "the best buildings are developed with the users."[28] As for the Beaumont Theater, there were no clearly defined users for the Koch Theater, referred to by the founders of Lincoln Center as the Dance Theater. Indecision about which organizations would perform there on a regular basis ruled out an orderly exchange between users and architect.

It was assumed that the New York City Ballet would occupy the Koch Theater, if only because Lincoln Kirstein, the preeminent patron of the ballet, was a member of the theater's exploratory committee. This brilliant, imperious intellectual was intimately associated with the New York City Ballet, having befriended George Balanchine and invited him in 1933 to come from Great Britain to the United States. The two men formed the Ballet Society in 1946 and the New York City Ballet in 1948, with Balanchine becoming director of the latter. In those days of autocratic decisions, Kirstein's single-handed designation of Philip Johnson as the architect went unquestioned, as did Johnson's similar designation of Vilhelm Jordan as his acoustical advisor.

Johnson was able to bypass the byzantine Lincoln Center system of multiple committee approvals thanks to Governor Nelson Rockefeller's ingenious suggestion that the 1964–65 New York World's Fair use the Koch for some of its events, thus enabling the state to make a hefty contribution to its cost ($15 million from the state compared with New York City's $2.5 million, the latter matched by Lincoln Center). With financing in place and no official occupant, the Koch Theater was the only building at the center to be completed on time and within one percent of its budget.[29]

The practical advantages of this simplified architect-client relationship were deceptive, however. What was gained in cost and timing was lost in meaningful communication with a client. Kirstein, who represented Balanchine, eventually told Johnson: "No one ever talked to George Balanchine." But according to Johnson, "Lincoln was not interested enough to follow the process of designing a building . . . So the result is that there are millions of things wrong with the theater."[30] Johnson was equally critical of the exterior.

New York City Ballet had no independent institutional existence, depending, like the New York City Opera, on the financial control and business arrangements of the City Center for Music and Dance on West Fifty-fifth Street, which the two companies shared. But inclusion of City Opera at Lincoln Center was violently opposed

View toward
auditorium, 1964.

by Rudolf Bing, the dictatorial, Viennese-born general manager of the Metropolitan
Opera, and by Bliss. Both men felt "the Peoples' Opera" was a competitive threat to
the Met. (And history has repeated itself: in 2001, Bing's eventual successor, Joseph
Volpe, just as adamantly opposed construction of a new theater at Lincoln Center for
City Opera.) Morton Baum, representing City Center, was equally intractable regard-
ing his organization's legal arrangements with Lincoln Center.

As usual following the directive of the Met, the Lincoln Center directors slighted
City Opera and, to supplement New York City Ballet's short season, thought up an
organization called the Music Theater for summer revivals of Broadway-style shows,
naming as director the composer and impresario Richard Rodgers (of Rodgers and
Hart and Rodgers and Hammerstein fame). By 1963 the Music Theater had become
a constituent and began its first season at the Koch in 1964; its instant and ongoing
deficit led to the operation's termination in 1970.[31]

Taking his cue from Rodgers, Johnson built the kind of small orchestra pit that
was common for the minimal group of musicians normally engaged for musicals
of that era (which has shrunk even further in the twenty-first century). Upon seeing
the space relegated to the orchestra during a visit to the building site, Balanchine
exploded in protest. Rockefeller's right-hand man, Edgar B. Young, recalled: "It wasn't
until I went round with Johnson and Balanchine and Kirstein after the theater was
nearly finished . . . that we realized that Balanchine simply could not visualize from

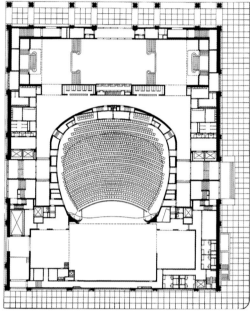

View toward stage,
1964; parterre plan.

blueprints, and until he stood on the stage and looked down into the pit, he hadn't realized it was not going to accommodate the size of the orchestra that they were already committed to have."[32] Kirstein and Balanchine's pleas to Nelson Rockefeller effected an enlargement of the pit—but it remained inadequate for opera.

In 1965 City Center became a Lincoln Center constituent, making its ballet and opera companies residents of the Koch Theater. When City Opera had its debut there that year it was clear that aspects of the 2,729-seat theater beyond the pit size were unsuitable. Johnson later admitted, "I didn't know what I was doing . . . It was absolutely marvelous to have a free hand. We knew we would have to have an opera orchestra, finally. See, nobody would tell me how many musicians. It seems unbelievable, doesn't it?"[33] As a result of this confusion—in the absence of a response to user needs—the building's prescriptive design doomed it to failure.

Alerted by *New York Times* music critic Schonberg to the theater's acoustical problems after a trial run there of the Metropolitan National Company, Julius Rudel, general director and principal conductor of City Opera, bravely retorted, "It won't be a problem; my singers are better than theirs." Despite his cool facade, Rudel, no longer trusting acoustician Jordan, immediately followed the brief telephone conversation with the critic by frantically calling J. Christopher Jaffe.[34] The conductor had already become aware of the musicians' difficulty in hearing one another during the ballet performances he had attended. For opera it was not so much the orchestra that concerned him but the projection of singing from the stage. Johnson's elegant proscenium arch was simply inside out—concave instead of convex—so that it reflected sound back to the stage instead of out into the audience.

The old rumor attributing the problem to a request by Balanchine that his dancers' footfalls be inaudible regardless of the overall acoustic consequences is denied by everyone associated with the project. George R. Steel, general manager and artistic director of City Opera since February 2009, summarily dismisses the idea, saying that it would be impossible to isolate this kind of noise. Johnson himself made the astonishing statement that "The acoustics didn't exist, because we were a dance theater" and deemed "hopeless" attempts to improve them.[35] What Balanchine did request was a stage floor resilient enough for the dancers' feet. The basketweave floor that fulfilled his wish, and set a precedent for similarly buoyant surfaces for dance, possibly muffled all sound on stage.

Before Rudel took up his baton at the theater, a number of small adjustments were made.[36] In his review of the opening night performance of City Opera at the Koch Theater, Schonberg—with minor caveats—was optimistic about the acoustics.[37] But as with Alice Tully Hall, though to a greater degree, time and experience changed perceptions. The Koch Theater's flaws—sound characteristics that mystified listeners with different experiences in different seats—were not helped by the ballet company's removal, for their season, of acoustical devices used by the opera company, and with time the discarding of these devices.[38] In 1981 the star soprano Beverly Sills,

Interior renovation, 1982; interior renovation, 2009.

who had become general director of City Opera in 1979, commissioned a $4 million acoustical renovation. It was completed the following year by Johnson and his then partner John Burgee, working with Cyril Harris, the expert who had consulted successfully for the Metropolitan Opera. Among other things, this overhaul permanently reversed the proscenium's shape, adding the so-called eyebrow above the orchestra pit, and filled in the porous ceiling that trapped sound above it.

In spite of all this, Paul Kellogg, who succeeded Rudel as director in 1996, described working in the theater as "an awful experience." He called in the Chicago acoustician Larry Kierkegaard, who installed an electronic enhancement system that provoked ongoing controversy.[39] City Opera officials defended this addition as "electronic architecture," pointing out that it was far subtler than the usual Broadway amplification. (In fact, similar systems have been used successfully in many venues; see Chapter 3.) Nevertheless, it was criticized by Sills, among others.[40]

The most recent overhaul, completed in the fall of 2009, was made possible by billionaire philanthropist David H. Koch, who pledged $100 million over ten years for renovation and for an operating and maintenance endowment. Together with his brother, Charles, Koch, an executive vice president and board member of a family conglomerate, has contributed to a variety of organizations.[41] The renovation has considerably improved the theater's strange acoustics. The work was initiated by the German company Müller-BBM and planned by JCJ Architecture with JaffeHolden and Arup as peer review acousticians. It cost $107 million of the $200 million capital campaign for the New York City Ballet and Opera that, until early 2011, shared the theater.

In planning the renovation, the acousticians had to be sure that their proposed improvements would not change audience perceptions of the dancers' footfalls. In order to do this Arup devised an enclosure, lined with a sound-absorbing treatment to approximate anechoic conditions and an array of movable microphones, with a slot on top for performers to enter. Recordings of the footfalls at different locations on the stage were used, via a computer model, to make sound renderings of the dancers' movements for the various refurbishment options.[42]

Several design changes directly affect the acoustics. The orchestra pit can now accommodate 105 musicians instead of 65; the pit has been placed on a mechanical lift in three sections so that the orchestra can play at any level, including—for concerts—flush with the stage (its curved apron straightened). An air plenum under the pit floor supplies the musicians with draft-free fresh air and allows them to hear each other better; it also gives the conductor better control of balances and provides the orchestra with presence without excessive volume.[43]

Lightweight partitions (which had hidden loudspeakers) at the stage ends of the balcony rings were taken down and replaced with sound-reflecting walls. At opposite

Foyer, opening night, 1964.

sides of the house, these floor-to-ceiling walls, perpendicular to the proscenium, enrich and carry into the hall the singers' voices and also improve hearing in the orchestra pit.

The velvety sound-damping carpeting that covered many of the theater's surfaces, as well as some walls, is gone, as is—most significantly—the controversial electronic architecture system. If the opening night performance of *Don Giovanni* in the 2009-10 season is an indication, the theater's new acoustics are encouraging. The sound of voices and instruments fills the theater in a straightforward manner, without the ricocheting effects that had troubled this space for so long. Once again, however, the complexities of judging acoustical quality were pointed up by *New York Times* critic Tommasini's varied experience. For the opening night gala, *American Voices*, Tommasini considered the orchestra's sound bright, rich, and with lingering resonance, but he felt that the voices lacked vibrancy.[44] At Hugo Weisgall's opera *Esther*, he commented that the acoustics seemed livelier, and while he gave high marks to the acoustics of the *Don Giovanni* performance, he noted the important role played by the stage set—jutting in part beyond the proscenium—in making the voices soar.

As suggested by Johnson himself, the Koch Theater had practical problems other than the acoustics. Apparently encouraged by Kirstein, the architect, acting alone, carved out the immense promenade lobby rather than providing adequate backstage space and acceptable working areas.[45] Faced with a choice between the two because of the cramped building site, Johnson—ever the elitist—didn't hesitate to short-change the theater's performers and managers in favor of its patrons. He later justified his decision: "Lincoln and I knew [Palais] Garnier, you see. We knew that the theater was the least important part."[46] Consequently, the rehearsal rooms for the orchestras in the basement and lower concourse are too small, and the offices that were left over for City Opera (after Balanchine had secured the only rooms with windows for the New York City Ballet) are subterranean cubbyholes—conditions that are not tolerated in theaters being built now.

On the other hand, Johnson's design of both the hall itself and the foyer—as he described it, "papered with people"[47]—was prescient. Ironically, what in the past was criticized—the inferior sight lines of seats on the balconies' sides and at the top—was preferred recently by David Taylor to the ski-slope side balconies and stepped side boxes of the 1970s that corrected this problem (but in his opinion drained theaters of their energy). For Taylor, people's movements in the balconies, as they shift in their seats for a better view of the stage, create the kind of interaction with the performers that is an attraction of live theater.

When applied to the arrangement of seats, however, Johnson's concept of "people papering" didn't work: his choice of continental seating has received as much criticism as the acoustics. That format succeeds for Alice Tully Hall's shorter parterre rows—as few as twenty seats and never more than forty-six. It was totally impractical for the Koch's fifty-six-seat rows, and a welcome feature of the current renovation is the insertion of an aisle within either side of that theater's parterre. According to George R. Steel, the new aisles also contribute to the acoustics, cutting pathways for the voices through the orchestra level.[48] The dialogue between the Koch's clients, architects, and technical advisors should provide sound equal in quality to the interior's fine vintage aesthetic.

Avery Fisher Hall

Formerly Philharmonic Hall
Max Abramovitz, 1962
Renovation, Foster + Partners, pending

Max Abramovitz based his ideal for the Lincoln Center concert hall on those he had visited in Europe.[49] In Vienna, Paris, and Amsterdam he was struck by the social interaction that took place in the theaters' large lobbies, where "people could communicate and walk and make it a functional thing." Contributing to this communal concept were eating facilities. Abramovitz got his generous circulation spaces without sacrificing support facilities (as Johnson had at the Koch Theater), but he was overruled on dining facilities, which were deemed inappropriate for a concert hall.

Opening night, September 1962.

The architect was criticized for replacing the red-and-gold color scheme usual for concert halls with blue walls and upholstery in light brown hues. (The auditorium's PR got off to a bad start when George Szell, then music director of the Cleveland Orchestra, asked for his opinion of the New York hall, responded, "Who can make music in a blue hall?")[50] Abramovitz's introduction of escalators to convey people up to the auditorium from the lobby was also distasteful to many. Yet it is Avery Fisher Hall's faulty acoustics that have received the most negative attention, from its opening night right to the present. It is difficult to think of any other concert hall that has been renovated, and once even entirely rebuilt, as many times as Avery Fisher Hall, and all this without ever receiving much approval of the acoustics.

Leo Beranek was the hall's chief acoustical engineer. His firm, Bolt, Beranek and Newman (BBN), enjoyed a fine reputation based principally on its work for industrial spaces, although by the mid-1950s it had contributed to several concert halls worldwide, including the Aula Magna Concert Hall designed by Carlos Raul Villanueva in Caracas, Venezuela (1953), and Eero Saarinen's Kresge Auditorium at MIT in Cambridge, Massachusetts (1955). Beranek had also visited music venues around the world to analyze their acoustical properties, a study he subsequently published in his 1962 *Music, Acoustics and Architecture*.[51] It was, however, BBN's work at the United Nations Headquarters with Harrison that landed the firm the job at Lincoln Center (shared with the English acoustician Hope Bagenal), where, in addition to Avery Fisher Hall, the firm was scheduled to advise the Metropolitan Opera, Alice Tully

Interior, 1962;
parterre plan.

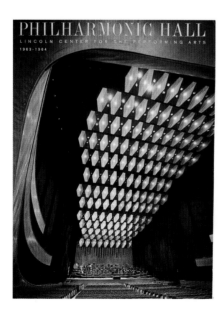

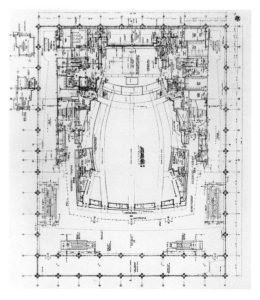

Hall, three Juilliard halls, and the Beaumont Theater. However, shortly into the Met project, the manager, Herman Krawitz, and Eric Leinsdorf, a conductor associated with the opera, became impatient with the tests being carried out at the old Met by BBN, so they chose instead the Danish acoustician Vilhelm Jordan, with Cyril Harris working on site.

Looking back at his experience with Avery Fisher Hall in 2009, the ninety-five-year-old Beranek said, "I should have resigned when they changed the seating from 2,400 to 2,750."[52] Both he and Abramovitz had based their work on the smaller number, on which the representatives of the orchestra had agreed. The adjustment in size, made independently by the Lincoln Center directors, is a measure of the lack of coordination between the project's participants. Carlos Moseley, the Philharmonic's associate managing director, was astonished to learn of the change in program upon returning to New York from a concert tour with the orchestra, as were Abramovitz and Beranek.[53] Interviewed in 2001, Beranek said: "We knew how to copy good halls, the same way a violin maker would know how to copy a Stradivarius. We did not know how to make a hall that was vastly dissimilar and that's where we got into trouble at Lincoln Center."[54]

Reviewing the inaugural concert in 1962 at Avery Fisher Hall, led by Leonard Bernstein, critic Schonberg complained of inconsistent sound: in the orchestra the sound was clear, somewhat dry, without much reverberation, and there was a lack of bass, whereas in the top terrace sound was more satisfactory.[55] One of the most troubling problems—as with the Koch Theater—was the inability of the members of the orchestra to hear each other. In another 1962 article discussing the auditorium's unsatisfactory acoustics, the New York Times's culture critic Arthur Gelb quoted Beranek as saying, "We're not going to tear down the hall and rebuild."[56] These words proved to be sadly prophetic: after Beranek's failed attempts to correct the acoustical problems quickly enough to suit the board, a long line of acousticians was brought in, starting in 1964, to try to remedy the situation, whereupon Beranek resigned from this and from the other Lincoln Center theaters.[57] Counting adjustments in 1963, 1964, 1965, and 1969, the cost came to more than $2 million.[58]

By 1969 the hall's fourth renovation was declared by Schonberg "a complete success" or, more vividly, "the greatest miracle since the birth of Tom Seaver"[59] (the baseball star). But even such enthusiasm from so influential a source could not stand up to the refusal in 1974 of both the Boston and the Philadelphia orchestras to perform there.

The year before, Avery Fisher, founder of the Fisher Radio Corp., had approached Amyas Ames, then chairman of both Lincoln Center and the Philharmonic, about naming possibilities; his $10 million donation met the price set for renaming Philharmonic Hall. Shortly after the name change, Fisher agreed that half of his gift

Interior, fourth
renovation, 1969.

would be used to rebuild the hall, spurring a comprehensive project by Philip Johnson
and Cyril Harris in 1976.[60] Remembering criticism of the acoustics at that time,
Fisher himself said years later, "Actually, the hall was not as bad as some people said
it was, nor was it as good as some claimed."[61]

For his Rug Concerts in the early 1970s, Pierre Boulez, the Philharmonic's
music director, achieved much better acoustics. To make an informal setting, rugs
and cushions were substituted for the seats and the stage was abandoned, with the
orchestra brought toward the hall's center, where it was surrounded by the listeners.
A large medallion was hung from the ceiling over the orchestra for lighting
and to make a new focal point. It was an extraordinary gesture. Boulez's wish to
attract a younger audience was fulfilled beyond anyone's most optimistic predic-
tions. Six concerts of contemporary music in a series of three seasons, with tickets
priced between $2.50 and $3.00 (moderate in comparison to the more usual $4.50
to $6.00 for Mostly Mozart tickets), were completely sold out. Furthermore, the
enterprise revealed vastly improved acoustics, implicitly criticizing the existing

architecture. Of course, moving the stage to the hall's center would have reduced the number of revenue-producing seats, but in addition to creating better acoustics, it would also have made the side balcony seats (currently at a right angle to the stage) point in the correct direction.

Encouraged by the experiment, Carlos Moseley began to work with the designer Peter Wexler to find a way of applying the arrangement to regular concerts.[62] It is intriguing to conjecture whether this investigation might have led to a simpler and less costly solution to the hall's acoustical problems than the complete rebuilding by Johnson and Harris, especially in view of the ongoing unhappiness with the results.

Philip Johnson's completion of the Koch Theater on time and on budget superseded for the board that theater's acoustical and other shortcomings, winning him the job for a new auditorium at Avery Fisher Hall. During construction, an interesting reaction to the plans came from Hans Fantel, whose column on high fidelity appeared regularly in the *New York Times*. Not only did Fantel point out the similarities between Harris's design and an early, unexecuted plan by Beranek (both were a shoebox shape, both had walls in a single plane with articulated facets, both had nonsloping, nonbulging balconies), he candidly stated that "the hall's troubles—at least at the start—stemmed not so much from acoustics as from politics."[63]

To document this opinion, Fantel referred to his colleague Schonberg's relentless railing against the sound at Philharmonic Hall: "Helped by the enormous influence of the *New York Times*, Schonberg made acoustical savvy something akin to a civic virtue and raised the temperature of the debate." True to Fantel's assessment of the

Rug Concert, 1974.

Rebuilt hall, 1976.

critics, after a brief assertion of the hall's improved acoustics, Schonberg pointed out that in contrast to the "warm, velvety sound" of older halls, Avery Fisher Hall's lesser reverberation produced a "cooler" sound: "In hi-fi terms Fisher Hall is 'flat,' without any peaks."[64] In other words, the hall was absent acoustic character or profile—essentially a neutral space, often preferred then.

Whatever imagined or real acoustical problems the rebuilt Avery Fisher Hall may have had, outstanding was the musicians' persistent inability to hear one another adequately at opposite sides of the stage; they were therefore unable to judge and effect balances. It was to address this problem that in 1992 rotund, basket-like sound baffles were installed high up on both sides of the stage.[65] Jon Deak, the Philharmonic's associate principal bassist, was among those who felt a dramatic improvement, despite some complaints that the brass had become too loud.[66]

Joseph Alessi, the orchestra's principal trombone, agrees with Deak's appraisal of the improvement of sound for the musicians while remaining critical of spotty sound in the auditorium. For Alessi, Avery Fisher Hall is "simply too large: the sound doesn't hold up."[67] But even more of a problem for the trombonist is the hall's appearance: "Other halls look glorious by comparison, and audience comfort is important, as are general facilities for the musicians that are missing at Avery Fisher Hall (such as practice rooms and warm-up spaces)."

As for Beranek, if he were presented with the Philharmonic Hall commission today, he says, "I would ask for more contact between the client, the architect, the acoustician, and the lighting experts." Not only did Abramovitz allow Beranek no direct communication with the client, insisting that all information come from the architect's office, he also made important design changes without consulting either the acousticians or Lincoln Center. One example was sloping the balconies upward from the front of the orchestra level instead of keeping them in their initial horizontal position.[68] Another was making rigid the suspended flat ceiling panels that were supposed to move in response to sound waves.[69]

Under the management of Zarin Mehta, appointed executive director of the New York Philharmonic in September 2000 (and president in 2004), together with a whole new generation of musicians and critics, there appeared for the first time greater satisfaction with Avery Fisher Hall. In a May 2009 article in the *New York Times,* Tommasini reviewed Lincoln Center's history: "The inadequacy of Avery Fisher Hall's current acoustics is greatly exaggerated. On a good night, when the Philharmonic is inspired, the sound has richness, clarity and presence. Did anyone in the audience think about acoustics in January when Gustavo Dudamel conducted the orchestra in an electrifying performance of Mahler's Fifth Symphony?"[70]

Sound baffles added, 1992.

Mostly Mozart Festival, 2006: audience on all sides of musicians.

In September 2009, upon becoming the Philharmonic's new music director, Alan Gilbert began to alter the musicians' seating, and he continues to do so. Every music director places the orchestra according to his or her own ideas and, with the musicians, begins the process of learning to "play" the concert hall. The son of two of the orchestra's violinists, Gilbert has long paid attention to the hall's acoustical variations. He literally grew up with them, even having had the advantage of hearing the results of various orchestral on-stage seatings that guest conductors can choose.

Most prominently, he split the violins left and right, rather than keeping them grouped together, as had often been the case before. And he moved the double basses and cellos from stage left to stage right. Also, Gilbert has not used risers, preferring to keep the musicians level on the stage floor. The opposite is true of visiting orchestras, which are universally praised for the quality of their sound in the hall, such as the London Symphony Orchestra and the Budapest Festival Orchestra: these ensembles place their musicians on risers of different heights, so that every tier of musicians can project sound into the audience without having the others as obstacles.

New Yorker music critic Alex Ross was the first to signal the favorable impact of Gilbert's rearrangements on Avery Fisher Hall's acoustics, describing "a more balanced, centered sound, with the hall's harsh acoustics tempered and the bass strengthened."[71] James R. Oestreich followed suit in the *New York Times,* writing that in its positioning, the London Symphony Orchestra led by Bernard Haitink "made Avery Fisher sound like a great hall."[72]

Considering these improvements, what does the Norman Foster/J. Christopher Jaffe renovation portend? Estimated in 2003 at nearly $326 million, the project (currently shrouded in secrecy) includes some design modifications that address the acoustics, among them surrounding the stage with the audience as it is for performances of the Mostly Mozart festival. The hall would be brought up to date aesthetically by reworking vertical circulation and public spaces.[73] In late 2011, no decision about what will be done was in place.

It is evident that perception of the auditorium as too large—too long for its width—has contributed more to this theater's failure than anyone is willing to admit. When the stage is in its original proscenium position, it appears far removed from the audience, causing a visual disconnect with the performers that is a factor in the continuing dissatisfaction with the hall's acoustics. What might take place to correct this condition is one of the many problems faced by the potential renovation.

Avery Fisher Hall may never rank among the great concert halls of the world—on a par with the venerable Musikverein in Vienna or the modern Suntory Hall in Tokyo, designed by Yasui Architects with Yasuhisa Toyota of Nagata Acoustics (1986). However, the recent reactions of music professionals hold out the hope that with careful usage, the auditorium can be an acceptable venue for the performance of classical music.

The hall is, however, handicapped by its prolonged, highly publicized acoustical problems, which may have produced audience prejudices reminiscent of the ones that greeted its opening. At that time, patrons of the Philharmonic, who had for generations sat in the same boxes at Carnegie Hall, were reluctant to accept a new theater from which the architect had stripped the boxes in order to bridge the separation between listeners and performers.

According to Zarin Mehta, the theater's problems are merely those of "an aging piece of real estate."[74] Adds Mehta, "This is one of the better halls we play in; it is not acoustically challenged." Perhaps a new renovation will at last satisfy the expectations with which the hall was inaugurated.

COMPARED WITH ITS APPEARANCE WHEN FIRST COMPLETED IN THE 1960S, Lincoln Center is a study in the updating of goals and the design process that is taking place at theaters worldwide. This revisionism is largely the product of a switch in attitude toward these institutions: from polite deference to restrained familiarity.

Rarely is the choice of an architect for such a theater made by a single individual—as it was by Kirstein with Johnson for the Koch. Equally unusual today is the exclusion of users from the design process—also the case at the Koch. The notion that client or architect can prescribe a building without extensive participation by users, acousticians, theater experts, engineers, and other technicians has mostly been discarded.

Before 1950 acousticians rarely participated in a building project. (Consultants are now often selected by the institution's board of directors before the architect.) And professional theater consulting is a relatively recent specialty whose degree of influence on design still needs to be fully defined.[75] Now the acoustician and the theater consultant, with the architect, comprise a triumvirate that is considered essential to the design of buildings for the performing arts. Acousticians participated in the original design of all the Lincoln Center theaters, but they in tandem with their peers in other technical professions continue to reach new levels of sophistication. Since 2000, modeling software has improved dramatically, so that designing spaces for sound is basically data-driven. At the same time, acousticians are increasingly using these computer models to create "auralizations" with which designers can listen to the model of the space and consider design variations. Much current research is focused on confirming the validity of auralization models and on refining the techniques by which they operate.[76] Finally, it is the communication between consultants, architects, clients (normally the board of directors or its representatives), and users that appears to outweigh in importance expertise in any single discipline.

The participatory design process parallels a similar trend in today's culture. The modernist concept of cultural "purity"—and a corresponding purity for the architecture serving it—died with its proponents. While early modernism eschewed the sacred, templelike architecture that characterized the new, democratic national museums as well as the large concert halls and opera houses of the nineteenth century, it continued to treat art and music as sacred.

The reversal of this attitude toward the arts has affected the architecture of the buildings housing them. The performing arts are no longer presented in elite bastions that hide mysterious activities. Elizabeth Diller's wish "to bring the Juilliard's activities from behind walls to the street" is emblematic. A new egalitarianism requires openness: for example, views from the street into the institution. Of the three main Lincoln Center theaters, Avery Fisher alone allows views of its lobbies from the sidewalk, a concept that has been much expanded by the new visual connection between Broadway and the lobby of Alice Tully Hall, as well as by a similar connection between Sixty-fifth Street and the Tully and Peter Jay Sharp Theater's lobbies.

Abramovitz's vision of a communal experience for concerts has become a goal for most art forms. Hugh Hardy, before his participation in the Lincoln Center renovation (a black-box theater for the roof of the Beaumont Theater), remarked of the original campus, "The fact that the entire complex was conceived without any food-related services except inside the Met would be inconceivable at this point in time."[77]

Site planning has also changed radically. People on foot, not automobiles, are now the focus of plans for the inner city. Historic plazas and dense commercial streets have become the model for pedestrian zones in city centers throughout Europe, and the trend is beginning to take hold in the United States. Planners have come to realize that super-blocks and broad, sunken plazas deferring to monumental buildings have deadened the street life essential to a lively urban environment.

In the United States and Europe, public institutions are currently woven into the cityscape to project a commanding presence. At Lincoln Center this is happening with Diller Scofidio + Renfro's more welcoming entranceways to the complex and attractions like the newly choreographed Revson Fountain, plus the increased porosity of the build-ings the firm has worked on. David Rockwell's contribution to the vitalization of the south side of Sixty-fifth Street is another gesture in this spirit. However, there is no getting around the fact that like most theaters, the backs of the houses at Lincoln Center make no connection with the streets they face.

Rather than siting several cultural institutions in one place, contemporary designs return to the earlier practice of locating concert halls and opera houses in different parts of a city—as the first Metropolitan Opera House and Carnegie Hall (William B. Tuthill, 1891) were in Manhattan, and the Royal Opera House (Edward Berry, 1858) and the Royal Albert Hall (Francis Fowke and H. Y. D. Scott, 1871) were in London. The extension of the campus's activities several blocks to the south with Jazz at Lincoln Center (Rafael Viñoly, 2004) at Columbus Circle is a step in this direction. New York's is not the only cultural center trying to overcome the restrictions of isolation: similar efforts are underway at the Dallas Arts District and elsewhere.

Even with its skillful updating, the Lincoln Center renovation is subject to anachronistic physical limitations. Chief among these is the size of its three largest venues—the David H. Koch Theater, Avery Fisher Hall, and the Metropolitan Opera House—which remains at odds with a trend in the 2000s toward smaller, more intimate theaters. In fact, the size of the Koch Theater has been a handicap to both resident companies—New York City Ballet and New York City Opera—a handicap that was proven insurmountable in May 2011 when City Opera departed.

The scale of these theaters thwarts the current desire to bring into new auditoriums a sense of the transparency that Diller Scofidio + Renfro achieved elsewhere at Lincoln Center with glazed facades. This interior transparency gives members of the audience more intense contact with the performers and a greater awareness of one another in scaled-down halls with seating in friendlier configurations.

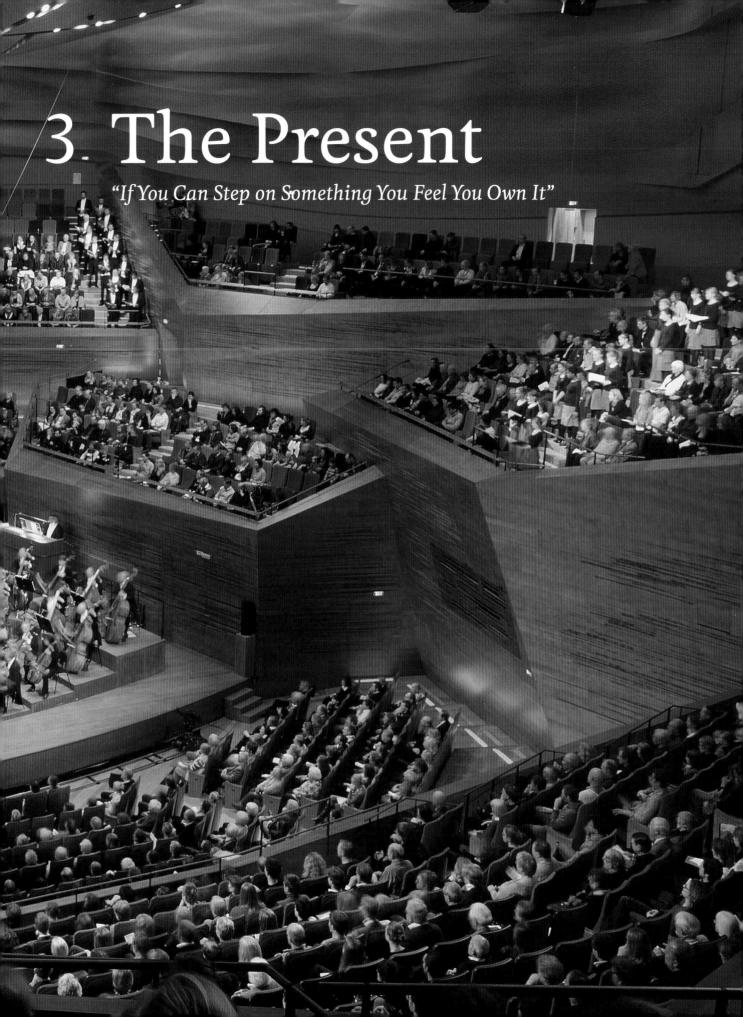

3. The Present

"If You Can Step on Something You Feel You Own It"

J UST PRIOR TO THE 2009 INAUGURATION OF OSLO'S NEW OPERA HOUSE, one of its architects said, of the accessible roof topography his firm designed, "If you can step on something, you feel you own it."[1] Recently, one of Lincoln Center's renovation architects expressed a similar idea when she described the center's grand new entranceway as "an electronic welcome mat."[2] The democratic approach of both statements is the outstanding characteristic of current attitudes toward the performing arts.

Such was not always the case. Today's approach harks back, in the United States, to the populist goals of the Works Progress Administration (WPA) in the 1930s, which reacted against elitist stances and tried to bring the arts to peoples' workplaces. (Unlike today, however, in an effort to make culture more familiar, the WPA encouraged the use of existing buildings rather than the construction of new ones.) In the 1960s and 1970s there was a return to exclusivity.

The cultural centers that began to rise across the nation in the 1960s were seen as temples of the arts: not only did they occupy new buildings erected for them, but the buildings returned stylistically to the grandiose neoclassicism characteristic of the late-nineteenth-century City Beautiful movement. The mid-twentieth-century proliferation of the performing arts contributed to an industry that in 2011 included 8,840 organizations and generated $13.6 billion in annual revenues (compared with $20.7 billion spent on sports admissions and $10.4 billion on movie tickets). A little over half of the revenue of not-for-profit performing arts groups is accounted for by theater and opera.[3]

The kind of architectural homogeneity that characterized Lincoln Center and so many other meccas of the arts in the United States (the Kennedy Center in Washington and the Music Center in Los Angeles, for instance) has been replaced today by eclecticism. Historic theaters in Europe—either preserved or re-created—are being supplemented by progressive architecture that is often more adventurous than its counterpart in the United States. If, however, today's theaters look very different from one another, they share certain traits: intimacy, transparency, and greater openness to the city are among the most important, replacing attempts to accommodate the largest possible audience.

Priority is given to making the public feel that it does indeed own whatever extravagant new theater has appeared in its midst. Almost every one of the opera houses and concert halls discussed here is located in a neglected urban area, and the intention is to improve those areas and attract other, less privileged audiences.

The participatory trend that has become paramount in the presentation of visual art applies also to the performing arts. While traditionalists cling to custom, forward-looking museum directors and music professionals are altering viewing and listening habits. Furthermore, audiences generally expect much more from a visit to the opera or concert hall—educational events, sophisticated food, socializing, shopping—than spectacle alone.

An important additional feature of the new venues is outdoor spaces. Practically without exception, the music venues in this chapter include exterior areas that are popular not only for promenading—the roof of Snøhetta's Oslo Opera House (2008) and the landscaped terraces of Frank Gehry's Walt Disney Concert Hall (2003) in Los Angeles come to mind—but in many cases for performances, which have moved outside (at OMA's 2005 Casa da Música in Porto, even the parking lot has been used for performances), recalling the theaters of antiquity as well as the nineteenth-century open-air concerts that took place in small parks that often adjoined concert halls of the time.[4] These outdoor events free performers and audiences from the expectations of perfect acoustics and formal behavior, favoring the social and urban aspects of the experience.

One characteristic of many large open-air concerts, such as those given in Chicago at the Jay Pritzker Pavilion in Millennium Park (2004) and in front of the New World Center in Miami Beach (2011)—both venues designed by Frank Gehry—is the use of electronic architecture (sometimes referred to as Electronic Reflective Energy Systems, or ERES).[5] This is not a standard reinforcement system but rather technology that creates the illusion of reflections, essential to the sound of traditional concert halls, in a room or outdoor environment where they would not otherwise occur. Developed from the 1930s on by early pioneers in the field, like Harry Olson of RCA, and greatly refined in recent years, these systems can simulate the sound heard in whatever environment is desired. Electronic architecture is used in many circumstances, as in the Metropolitan Opera's high-definition broadcasts, for which state-of-the-art sound systems that approximate the acoustical quality of the Lincoln Center house are installed in the theaters being used.

There have been numerous successes with these systems, exemplified by the Hult Center for the Performing Arts (1982) in Eugene, Oregon, and the Hilbert Circle Theater (1984) in Indianapolis, Indiana. But they are designed primarily for classical music, not for use with amplified programs, and are therefore not equally appropriate for all kinds of performance.

Although electronic architecture can come surprisingly close to the real thing, the architects and acousticians of the following buildings for music have gone to extraordinary lengths to provide first-rate natural acoustics as well as visual appeal. These new theaters also offer typical sound-reinforcement systems for popular programs, and in a few cases electronic enhancement.

There is additionally a new willingness to experiment with materials, notably glass, that were formerly considered acoustically inappropriate for concert halls. To overcome the acoustic drawbacks of smooth glass, for example, architects have devised novel ways to install glass panels and inventive means of fabrication to shape the material itself. In several cases, inherited assumptions that traditional interiors assure superior acoustics have discouraged architectural experimentation. Yet many exteriors are so arresting that the conservatism of their theater interiors can be over-looked. The Oslo Opera House stands out in this category, as does the Grand Canal Theatre (2010), Dublin, by Daniel Libeskind, though to a lesser extent. Norman Foster's Margot and Bill Winspear Opera House in Dallas, Texas (2009), provides excellent acoustics in an elegant setting, but its most interesting feature is its outdoor theater.

For the most part, recent concert halls are more inventive architecturally than new opera houses. The complex hybrid of opera—music, theater, often dance, and increasingly video—needs the qualities of a good concert hall for symphonic music as well as those of a good drama theater. When the theater is a multiuse venue in which concerts of symphonic music are also performed, the acoustical program becomes even more complicated. Finally, and most tellingly, to survive financially, opera houses must be able to share productions, and therefore cannot be so different from one another as to exclude this possibility.

Conversely, some new interiors are remarkably imaginative. The Dee and Charles Wyly Theatre in Dallas (2009) by REX/OMA is based on the ingenious concept of stacking support spaces vertically. While by no means an opera house in the usual sense, it has the ability to accommodate diverse musical presentations in a unique, reconfigurable setting.

In contrast to the typical opera house (with fly tower and elaborate backstage machinery), in which the stage and audience are separated by a proscenium, the concert hall is usually a single, unified space that can dissolve the distinction between illusion and reality. Frank Gehry told the students in his Yale School of Architecture studio class that he considers one-room buildings "powerful because there is no place to hide."[6] The statement applies to the in-the-round seating of the architect's Disney Concert Hall and to his New World Center concert hall in Miami Beach. Gehry's multiuse proscenium theater in the Richard B. Fisher Center for the Performing Arts at Bard College in Annandale-on-Hudson, New York (2003), is a different type of theater.

The Muziekgebouw aan't IJ/Bimhuis in Amsterdam (2004) by Danish architects 3XN, Rem Koolhaas's Casa da Música in Porto, Portugal (2005), and Jean Nouvel's Danish Radio Concert Hall in Copenhagen (2009) each bring a distinctiveness to concert hall design. In addition to the unusual interiors of twisted oak, the Royal Conservatory of Music's Koerner Hall in Toronto (2009) by Marianne McKenna (of Kuwabara Payne McKenna Blumberg Architects) offers a conservative model for expanding an existing facility.

Three halls in Austria provide the opportunity to hear music better suited to restricted spaces, as well as challenging works for which there is a limited audience. In the United States, even the bigger theaters of the late twentieth and early twenty-first centuries have fewer seats than comparable ones of earlier generations. Compare the 2,265 seats of the largest discussed here, Disney Hall, with Avery Fisher Hall's 2,750; and the Winspear Opera House's 2,200 is more in line with houses in Europe, where smaller is the rule. There are several reasons for the downsizing trend. For one thing, it is easier to achieve acoustical excellence and intimacy in smaller spaces. For another, musicians prefer a full house to a half-empty one.

The Haus für Musik und Musiktheater (MUMUTH; 2009) in Graz, by UNStudio, includes an auditorium whose natural acoustics can be altered by electronic architecture. The Glass Hall (2004) in Vienna by Wilhelm Holzbauer and the Crystal Room (2009) in Waidhofen an der Ybbs by Hans Hollein give intimacy a new meaning: both offer excellent sound in agreeable settings. Like the first purpose-built concert halls, these three mini-halls all have fewer than five hundred seats.

Zaha Hadid characteristically brings exhilarating forms to the performance of music. Her JS Bach Chamber Music Hall (2009) is a movable metal-and-fabric structure that expands the concepts that inspired temporary twentieth-century pavilions for which specific works were composed. In addition to its ethereal environment, the pavilion's minimal impact on acoustics, which depend on the existing space in which it is installed, demonstrates the power of psychoacoustics.

Oslo Opera House

Oslo, Norway
Snøhetta, 2008
415,410 square feet
$750 million

No existing or projected building for classical music does more to dispel hallowed historical associations than the opera house in Oslo, whose roof has become the city's favorite new promenade. An enormous ramp rises steeply from the waters of the Oslo Fjord, appearing to extend to infinity due to the absence of visible rails or balustrades at the top. The slope creates a new topography for this area of the city.

Related to the concept of "oblique architecture"—a means of social transformation advocated by the architect Claude Parent in the 1960s—the opera roof is animated day and night by young and old, dogs, baby strollers, and skateboarders undeterred by a dramatically steep incline and potentially perilous incisions and uplifts. It is indeed, as principal Craig Dykers says, architecture "you can step on."

Thrusting skyward within this momentous landscape and carpeted in nearly two million square feet of Carrara marble is the glazed volume of the foyer and, upright behind it, the fly tower, which is sheathed in aluminum panels. The 415,410-square-foot building sits partially on pylons sunk nearly 200 feet underwater (at tremendous expense) and partially on the shore, not far from downtown Oslo. Seen from the city the opera house appears as a series of powerful zigzags: diagonal layers formed by the stone-clad plinth, the glazed main entrance, and the glazed foyer top.

A relatively narrow footbridge over a small canal leads from the mainland to a vast plaza that flows along both sides of the opera house to the roof ramp. Glass walls allow ample views into the airy foyer, which rises in places to sixty-five feet. Facing the foyer, the exterior of the auditorium—a circular form reminiscent of New York's Guggenheim Museum—is sliced open at each balcony level. The surface consists of thousands of different sized oak slats acoustically designed to diffuse sound within the foyer. The same honey-colored oak also sheathes the walls of the curved corridors leading into the theater. Tapered concrete columns cant to support the foyer's lofty roof. Tucked into one of the sloping spaces resulting from the roof's unusual shape are three detached cement cubes designed by the artist Olafur Eliasson. Art objects that contain the lavatories, the cubes have perforated and patterned skin lit with changing painted LED-MDF lights.

The conventional three-tiered horseshoe of the 1,370-seat main theater is a letdown in comparison to the exciting exterior. Dykers says that from the beginning,

the theater was to be traditional: "I didn't want to reinvent this kind of room. It isn't for experimental theater. It had to be well-protected and a particular shape." That shape was specified by Theatre Projects Consultants, with whom the architect has worked for twenty-five years (and conformed to the client's wishes).

The warm blond oak of the theater's foyer walls and corridors appears dour when stained darker for the interior walls and balustrades. Seating is also less democratic than the popular roof promenade might lead operagoers to expect. The most affordable uppermost seats are cramped (recalling comparable seats at the Palais Garnier) and, like other balcony locations, suffer from the presence of intrusive

Lobby; theater.

lighting fixtures. The large pit disconnects the audience from the stage. The acoustics (by Brekke & Strand Akustikk and Arup Acoustics) are bright if uninspiring.

Dykers points out that a traditional room within a contemporary design is typical of Snøhetta's intentional dualities. Another such contrast, he says, takes place between the building's marble, glass, and steel public areas—"a city of imagination"—and the single-level, aluminum-covered production area—"the factory"—that reflects the heavily industrialized nature of the developing Bjørvika district it faces. Separated from the public part of the building by a fire and acoustical barrier and a wide service corridor, the boxlike structure provides six hundred people with offices, workshops, rehearsal halls, and a ballet academy. These facilities, most of which are visible from the exterior, are grouped around a large, open-air garden courtyard reserved for staff and performers.

A 190-seat black box is for spoken drama, and the 440-seat Stage 2 is intended for experimental opera and dance. Regrettably, the acoustics in the latter are disappointing. Members of the cast of Georg Friedrich Haas's opera *Mélancholie* reported that they needed to strain their voices to project. And Tom Remlov, the opera house's director, repeats the common complaint that, in addition to other problems, Stage 2 is too expensive to reconfigure.[7]

Remlov's complaints are few, however; for the most part he is thrilled with the building. Instead of using the flawed Stage 2 as an alternative to the main theater, he prefers the building's exterior: the large flat area at the roof's northernmost side is amenable to chamber music concerts, for example. Inside, the generous, day-lit rehearsal room in the back of the house can also be used for public concerts.

Snøhetta was founded in Oslo in 1989 by five young architects for the purpose of entering a competition for a new library in Alexandria, Egypt; the firm won the competition and the library was completed in 2001. The Oslo Opera House, the firm's next world-class commission, is a tour de force in its combination of monumentality and accessibility.

Plan.

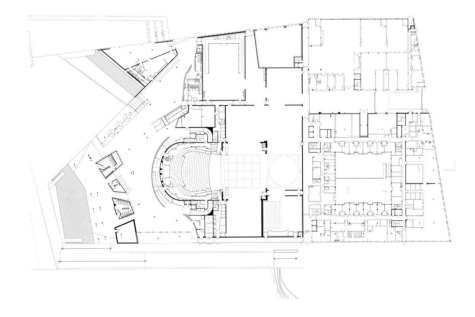

Margot and Bill Winspear Opera House

Dallas, Texas
Foster + Partners with Kendall/Heaton Associates, 2009
234,400 square feet
$150 million (approximate)

The entrance to the Margot and Bill Winspear Opera House is overwhelmed by the massive solar canopy (463 by 378 feet) surrounding the building. The steel canopy, with its variously angled anodized aluminum louvers, extends the structure's reach outdoors and helps to define an enlarged public area, but its scale and rigidity are oppressive. A lighter touch might have benefited the design of this portico, which gives solar protection to a public plaza and to the sixty-foot-high glass walls that wrap around the lobby. In good weather, sliding glass panels that run the eighty-four-foot width of the lobby's east side open so that patrons of the Winspear's restaurant and café can sit or mingle outside. By providing a temperate outdoor oasis, the canopy reduces heating and cooling loads on indoor spaces, one of a number of energy-saving systems including, in the hall itself, a displacement ventilation system (a common feature in new theaters) that pumps air from the floor so that it cools the audience but not the space above. The architects landscaped the shady refuge under the portico with squares of lawn and wildflowers and a black-granite reflecting pool whose film of water is flush with the surrounding pavement.

Norman Foster and his partner Spencer de Grey's use of red on the outside instead of the inside makes the opera house the Dallas Arts District's most prominent constituent. Foster has a knack for creating unusual theater architecture. The Sage Gateshead (2004) in northern England, his first essay in this building type, is a wavy, steel-and-glass behemoth that has been compared to a "moored blimp."[8] Two years after completion of the Sage came the Palace of Peace and Reconciliation in Astana, Kazakhstan, the new, Brasília-type capital laid out by the Japanese architect Kisho Kurokawa. The pyramidal form of Foster's building is arguably the weirdest opera house in existence. In fairness to the architect, the 1,500-seat auditorium was a latecomer to the program, which originally called only for a meeting hall and conference rooms.

In Dallas the architects covered the opera house's concrete drum (which contains the auditorium, stage, fly tower, and cooling towers) with bright ruby PVB (polyvinyl butyral) laminated between two sheets of glass. The colorful skin is illuminated from both back and front, making bold signage possible through the day and especially at night, when the building is washed in a red glow of light.

Exterior view.

Palace of Peace and Reconciliation, Astana, Kazakhstan, 2006.

For all the exterior panache of the opera house, its entranceway is singularly undramatic. The architects dared, in this car-dominated city, to locate elevator and escalator egress from the underground parking outside the structure rather than within, thereby forcing all patrons to enter through the same set of doors. But because those doors align with a new north-south path perpendicular to the arts district's main axis, Flora Street, instead of with the grand stairway, visitors enter to face a wall. The interior's red-glass panels, so effective from the exterior, fail to dispel the corporate feeling of the glass-and-aluminum lobby. Circulation, nevertheless, is efficient, via elevator and wide stairs that hug the theater's curve, with bars and cafés on the first and second levels.

Bill Winspear's $43 million donation to the Dallas Opera House gave him a major role in the project—both in selecting the architect and acoustician and in establishing exactly what their goals should be. Robert Essert, founder of Sound Space Design and the chosen acoustician, recalls the Canadian-born entrepreneur's insistence that "the number-one priorities should be excellent acoustics and sharing his passion for music with as many people as possible."[9]

Winspear's insistence that the new house be first and foremost for opera, with other kinds of performance such as dance and touring shows relegated to second place, gave Essert his directive and determined many of the architects' decisions.

Lobby, with grand stair; plan.

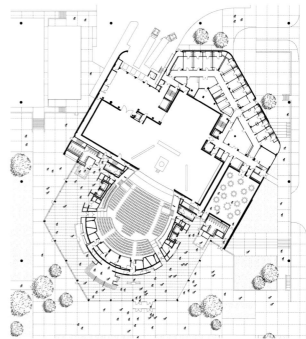

Theater.

Essert says, "Bill Winspear didn't actually specify what *kind* of sound he wanted. He
wanted acoustical excellence for opera that would be on a par with the [Morton H.
Meyerson Symphony Center] acoustics next door. The Dallas Opera does mostly core
opera repertory (from Mozart to Strauss). So it was important to make a warm sound
for those orchestras, with decent resonance." Graeme Jenkins, music director at
the time, concurred with this viewpoint, using the lush, dark orchestral sound of the
Vienna State Opera as a reference point.

    With these guidelines in mind, Essert approved horseshoe seating for the theater—
this, in his opinion, was "a guarantee of good acoustics."[10] That he achieved this and
more is attested by *New York Times* music critic Anthony Tommasini, who cited the Winspear's exceptional combination of "richness and resonance" in addition to its bright,
clear sound.[11] In my own experience there, a performance of Verdi's *Otello* provided
well-balanced orchestra sound for seats both in the parterre level and the top balcony.

    The illuminated silvery, crinkled fronts of four gently sloped tiers of balconies
stand out against the dark brown peripheral walls, defining the interior space and
making the 2,200-seat theater feel smaller. Continuation of the top tier of the balconies (though without seats) in a ring in front of the barely noticeable proscenium
contributes to this effect.

A fire curtain designed by the Argentine artist Guillermo Kuitca protects the deep stage, which has an ample backstage and wings, all with state-of-the-art technology. The theater boasts a generously scaled rehearsal room; set design and storage are off site. Charcoal-gray Ultrasuede upholstery, burnt-walnut floors, and a spectacular seventy-foot-tall retractable chandelier made of 318 long, thin acrylic light rods complete the house's elegant decor.

De Grey points out that many of the architectural decisions were made with acoustics in mind. For the most part, the building consists of timber on concrete; in the theater, plaster walls on masonry reinforce bass response, and slightly convex peripheral walls help disperse sound and prevent echoes. A large, open orchestra pit that can be raised and lowered accommodates up to one hundred musicians.

The Winspear stands in a long line of opera houses that have reserved innovation for the exteriors rather than for the theaters themselves. The Dallas opera house, like the Oslo Opera House, recalls the strictly symmetrical horseshoe configuration established by Carlo Fontana in Venice in the late 1600s.

By the time planning for a Dallas cultural center began in 1979, the sixty-eight acres designated were already cut off from downtown by freeways to the north and south, creating a desolate no man's land. The 1956 National Interstate and Defense Highways Act had made this a common occurrence in many American cities. Dallas is now struggling to reconnect the arts district with the city, just as Lincoln Center is doing. But in Dallas the task is more demanding, involving the construction of a 5.2-acre bridge, at an estimated cost of over $114 million, across the Woodall Rodgers Freeway. In an effort to attract people from the residential community to the north, this so-called deck park will include a playground, dog park, botanical garden, picnic area, and restaurants.

Linking the cultural center with the city also called for revisions to the original master plan, which had rigidly aligned various institutions along a single thoroughfare: the Dallas Museum of Art (Edward Larrabee Barnes, 1984), the Meyerson Symphony Center (I. M. Pei, 1989), and so on. Foster and the architects of the Wyly Theatre across the street from the opera house—Rem Koolhaas and Joshua Prince-Ramus—enlivened the plan with new axes that link their buildings to the surrounding roads and to plazas intended to attract non-theatergoers. Consequently, the Winspear now shares with the Meyerson a ten-acre park and an orientation rotated thirty degrees from the street.

De Grey reserves particular enthusiasm for the outdoor performance space that is intended to reinforce the building's image of accessibility. Adjoining the opera house and protected by the solar canopy, this open-air square will continue a local tradition of pop concerts and fiestas that for decades has attracted as many as five thousand people a night.

Dee and Charles Wyly Theatre

Dallas, Texas
REX/OMA (Joshua Prince-Ramus and Rem Koolhaas), 2009
80,300 square feet
$354 million (approximate)

Like everything about the Wyly, the way in which the theater reaches out to its surroundings is unusual. The very heart of the design by Koolhaas, the founder and
principal partner of OMA/AMO, and Prince-Ramus comes from the decision to expose
the 575-seat theater to the exterior. Wanting to overcome the conventional conceal-
ment of the auditorium, the architects stacked the support spaces—which usually
wrap the stage house—in ten vertical stories and buried the lobby underground.
Consequently, the twenty-eight-foot-high curtain wall of sound-diffusing glass, with
integral shade controls, on three sides of the theater reveals what is going on inside,
besides affording views out. Two windows (cut back from four by budget considerations) pivot open to the exterior.

Interior view, with
blinds up.

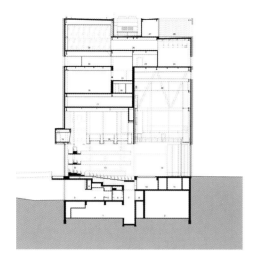

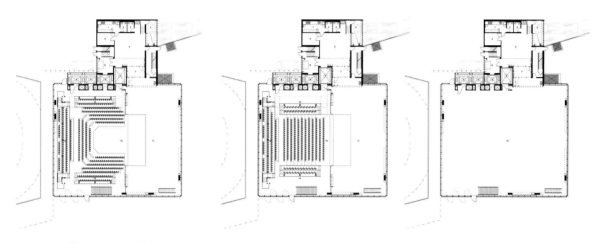

For all its ground-floor transparency, the twelve-story Wyly is an enigmatic presence across from the Winspear and diagonal from the Meyerson. Its verticality contrasts with the low-slung opera house and the symphony center, relating rather to the office towers behind. The theater's nearly windowless, tubular-aluminum facade, which hovers above the porous street level, is much more interesting than Foster and Pei's staid buildings.

As at the Winspear, however, the entranceway to the Wyly leaves something to be desired. In an inversion of the historic grand staircase going up, a ramp and driveway go down a broad landscaped area in which lighting wedges are embedded. The proximity of cars and pedestrians is unfortunate, if not dangerous.

And problems with circulation continue inside the building. The low-ceilinged, reinforced-concrete lobby, paneled in stainless steel, and the narrow, chain-mail staircase leading to the theater make an effectively compressed prelude to the latter's expansiveness. But having to walk down to the lobby and then up to the theater (and back down to the lobby for the toilets and concessions) has been difficult for older theatergoers and those with physical impairments. (An elevator is generally ignored.) Still, this has been the only problem posed by the design's innovative verticality, according to Kevin Moriarty, the Dallas Theater Center's artistic director at the

Exterior view; section; plans.

Wyly.[12] Otherwise, Moriarty reports, the stacking of workshops, offices, and rehearsal spaces is "unbelievably user-friendly."

Freeing up the ground plane from columns has given unprecedented flexibility to the auditorium. In fact, the Wyly expands the kind of workable modular theater that architects have been trying for since the first part of the twentieth century. The Dallas theater goes well beyond Walter Gropius's idea of interchangeability between proscenium, thrust, and flat-floor arena (see Chapter 1) in its ability to reconfigure in record time and with cost effectiveness. Walls, scenery, and floors are adjustable electronically, and balcony seat towers can be retracted vertically within a day by three to four workers.

The acoustician Renz L. C. J. van Luxemburg (architectural acoustics, Eindhoven University of Technology) applied sound-absorptive panels to the ceiling above the technical grid for conditions conceived primarily for drama. However, Koolhaas says that myriad spectacles can be performed at the Wyly: "I had everything in mind, including various kinds of music. It was meant to be an anti-dogmatic theater to show that anything can happen anywhere."[13]

Moriarty concurs. "The theater," he says, "has revealed itself to be everything Rem and Joshua had hoped for conceptually." Because one of the two Broadway musicals (*Superman*) staged during the first nine months of operation uncharacteristically required no amplification whatsoever, Moriarty feels that classical music could be performed in the same manner. Among the professionals who look with envy at the Wyly, George R. Steel, director of the New York City Opera, says, "It would be great fun to do opera here."[14]

Grand Canal Theatre

Dublin, Ireland
Studio Daniel Libeskind, 2010
148,171 square feet
$102 million

Daniel Libeskind's poignant evocation of the Holocaust in the Jewish Museum Berlin (1999) earned him instant fame and launched a career that had until that time been primarily in academia. He now heads offices in Manhattan, Milan, and Zurich. But Libeskind asserts that it was his brief, earlier life as a professional accordion player that influenced his approach to the Dublin project, his first theater design: "Having been on the stage myself," he says, "I understood the audience's point of view, but also the performer's."[15]

Since 1997 the self-financing state body in charge of the Dublin Docklands Development Authority has overseen construction of commercial and residential

Exterior view.

buildings to replace the derelict shipping and industrial facilities on the north and south banks of the River Liffey, celebrated in James Joyce's *Finnegans Wake*. Connecting with the south side of the Liffey is the picturesque Grand Canal, where Libeskind's theater and plaza are located. They signal a major cultural presence, a welcome relief from the area's anonymous commercial architecture (including over half a million square feet also designed by Studio Daniel Libeskind).

To acknowledge the waterside site, the architect wanted to use nautical imagery throughout the theater interior, as in his acoustical sail-like forms that float beneath the ceiling. Other aspects of this idea were to have been an asymmetrical space with a freer, more casual seating arrangement and a more modern relationship between the audience and the stage. This concept fell by the wayside at the insistence of the de facto client—Harry Crosbie, impresario to superstars, of the huge worldwide, California-based events company Live Nation (owner and operator of the theater). Like the Kunsthaus museum typology, which is without an art collection of its own, this touring house stages only productions that originate elsewhere, primarily Broadway-style musicals.

The architect claims not to have been inhibited by his client's strong guidelines. On the contrary, "Crosbie's very clear idea of what he wanted, even with the restrictions this entailed, was a huge imaginative stimulus," he says. "I had a very different index of freedom from Scharoun, who was dealing with a new city. I had to meet classic criteria."

Theater.

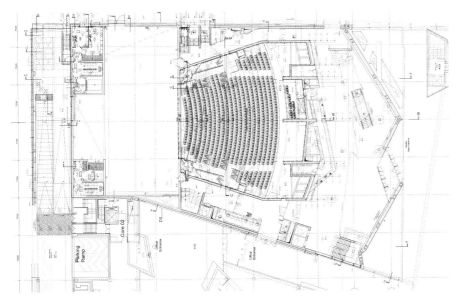

The result is a conventional, fan-shaped auditorium for 2,100 with two central
balconies and narrow boxes at each side; it was modeled on the prototypical English
music theater developed by the prolific, self-taught architect Frank Matcham
around the turn of the twentieth century.[16] The same precedent dictated the red color
scheme (in four different hues) for the walls, seats, and all-over carpeting, with walnut-
veneered doors and touches of gold on the balconies.

In contrast with most opera houses, designed primarily for use without amplification, the Grand Canal Theatre's priority is amplified music. It nevertheless offers
the best facilities in Dublin for large-scale opera. According to Michael Dervan,
the music critic of the *Irish Times*, natural acoustics (by Arup Acoustics, UK) were
adequate for the June 2010 production of *La Bohème*, part of the ten percent of programming the house devotes to classical music annually. Dervan wrote that his
impression was "of sound . . . that was clear and on the dry side. It certainly wasn't
either warm or full. But it was perfectly viable."[17]

Libeskind compensated for the enforced banality of the auditorium with the
dramatic expressivity of the building containing it. Inspired by the wedge-shaped
stage sets he designed for the opera *Intolleranza* by Luigi Nono at the Staatstheater in

Plan; section.

Saarbrücken, Germany, in 2004, the architect used an expanded wedge for the theater's reinforced-concrete and steel structure. Within the pointed front, "the beak," is a soaring space that rises the full six levels of the building and cantilevers over the plaza.

Enormous angled and canted panels make up the glazed facade, behind which tall, leaning structural supports and the loud red exterior of the auditorium are visible. The effect, especially when illuminated after dark, is of a giant proscenium with the folds of a curtain drawn closed within it.

Also working in a theatrical analogy, the landscape architect Martha Schwartz used resin-glass pavers for a "red carpet" that extends from the edge of Dublin Bay to the entrance portal. In her usual allusive manner, Schwartz chose to reference the Italian Grand Canal, rather than the one the theater faces, with a series of tilted, bright red light sticks modeled on Venetian mooring poles bordering the pathway. She populated the rest of the plaza with polygonal elevated planters filled with wildflowers. Scattered among the planters are apple-green perforated-metal benches that invite strollers to sit and enjoy views of the bay. At one side of the entranceway, the jagged flat slabs of a large fountain are an awkward approximation of Libeskind's splintered forms.

Libeskind regards the plaza as an outdoor lobby, and in fact the actual, interior lobby is narrow (as those of Broadway theaters are, for economy's sake), its height compensating for the cramped width at ground level. At the entrance level and each of the four floors above, a generous refreshment bar borders the auditorium's outer periphery. On the wider second level, a pub is part of an elegant club lounge reserved for Dress Circle and other privileged patrons. Every one of these atrium spaces enjoys magnificent views of the water and its surroundings.

Together the plaza and the theater facade create a contemporary *architecture parlante:* a VIP red carpet leading from the bay to a giant stage front. The scenario cries out for the possibility of arriving by boat, as the public can, for example, at the Oslo Opera House.

Walt Disney Concert Hall

Los Angeles, California
Gehry Partners, 2003
240,000 square feet
$207 million

Now that Disney Hall is firmly positioned as a major player in the international music scene, it is easy to overlook its difficult beginnings. In 1987, when Gehry won the competition for the hall, Frank O. Gehry & Associates numbered about thirty people, and computer-aided design was in its infancy. The Disney was Gehry's first major job in his adopted hometown of Los Angeles; he had not yet won the Pritzker Prize (awarded to him in 1989), and the small Basque city of Bilbao was largely unknown.

With little confidence in what was considered at the time an inexperienced team, the multiple clients—the Los Angeles Philharmonic, Los Angeles County, the Disney

Exterior view.

family, and the Concert Hall committee—required that another, more established firm produce construction documents. Dworsky Associates' failure to properly interpret the Gehry design shut down the project in 1994. By the time it started up again in 1997 with a new infusion of funds, Gehry's professional life had taken a quantum leap forward.[18]

Since 1991, Gehry and his associate James Glymph had been using the CATIA three-dimensional computer-modeling program developed by Dassault Systèmes for the French aerospace industry. Gehry was employing it not for design but for fabrication of the infinitely complex sculptural forms that his work had assumed. It was this system of digitizing traditional designs and models that enabled construction of the Bilbao Guggenheim; its stunning success in promoting the depressed Spanish city positioned Gehry as the most famous architect in the world. Indeed, "Bilbao effect" entered common parlance to describe the mythical ability of high-profile architecture to jump-start urban renewal and put a city on the map. By the time the Basque museum opened, what had become Gehry Partners numbered about 130; it now employs more than 200.

Buoyed by his triumph, Gehry was in a strong position to retake control of the Los Angeles project, which had already evolved into something very different from his first design. The French acoustician who had consulted for the competition had been superseded by a Japanese firm headed by Minoru Nagata, who, upon his retirement, was followed by his protégé Yasuhisa Toyota. Under the new acoustician's guidance, a traditional shoebox shape replaced the segmental plan intended by the architect to evoke a floral image, a homage to Walt Disney's widow, Lillian Disney, the hall's chief patron. For budgetary reasons, stainless steel was used instead of the French limestone sheathing originally selected, a decision Gehry still regrets since he had anticipated the metal's much-criticized harsh solar reflections.[19]

Competition plan; realized plan.

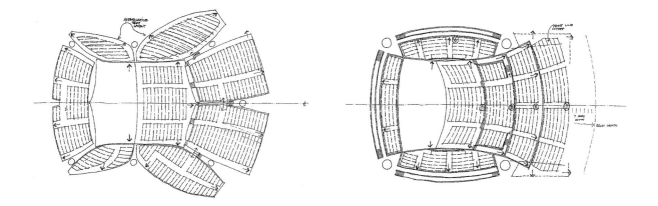

Auditorium.

Like the Dallas Arts District, downtown Los Angeles is cut off from the rest of the city by freeways. Disney Hall's combination of exuberance and friendliness makes it a celebratory oasis in what was for decades a lifeless corporate area. Everything about the building is inviting. The exterior's billowing stainless-steel panels protect the broad travertine stairs flowing into the tall, light-filled lobby, where the glass wall can be raised to open the lower level to the sidewalk. Punctuated by giant treelike columns, the lobby contains a café, gift shop, and the stunning Founder's Room. The space hosts lunchtime concerts and lectures and has ample views to the exterior and access to the generously landscaped public terraces (designed by Melinda Taylor and Lawrence Moline). As for the foyer, the terraces are open to non-ticket holders.

In the main hall, precedence was given to the performance of classical music. Gehry was determined to avoid the movable ceiling canopies and adjustable acoustic chambers that for the most part went unused in many auditoriums (I. M. Pei's Meyerson Symphony Center, for one). His decisions were encouraged by, among others, Ernest Fleischmann, managing director of the Los Angeles Philharmonic (1969–98), and Esa-Pekka Salonen, who became music director in 1992.

Unlike Gehry's two subsequent concert halls, the Fisher Center and the New World Center, the Disney Hall stage, placed within audience seating, remains basically fixed. Locating musicians offstage was also ruled out. Experimentation is possible, however, in the transformable 180- to 280-seat black-box REDCAT (Roy and Edna Disney/CalArts Theater) auditorium and gallery, which is tucked into the underground garage area.

The walls and ceiling of the 2,265-seat auditorium are mostly of Douglas fir, its floor is red oak, and like Vienna's famed Musikverein and other nineteenth-century

halls, it enjoys natural light. At Disney, daylight enters through a generous window and skylights, cheerfully animating the interior. The festive mood is reinforced by the brightly colored floral pattern of the seating upholstery, a nod to the botanical interests of Mrs. Disney and a cogent reminder of the southern California environment.

Reinforcing the visual warmth of the timber interior are the graceful bowed segments within which the hall's seats are distributed. Seating is at different levels and in different arrangements in front of the stage, on each side of it, and behind it, following Scharoun's example as well as the arrangement that contributed to the acoustical excellence established by Toyota at Suntory Hall in Tokyo (1986). Most numerous are the seats in the steeply raked orchestra level. The explosive forms of the organ, designed by Gehry, dominate the hall.

Gehry was able to maintain his distinctive sculptural style within the two-layered rectangular box dictated by Toyota, who also specified the room's height (which decreases from sixty to about fifty feet above the conductor's platform). The ceiling's slightly curved interwoven strips of wood veneer are like draperies behind and at the sides of the stage.

Critics who have praised the acoustics of the Berlin Philharmonie and Disney Hall refer, in both cases, to the phenomenon of psychoacoustics. By conventional standards, neither hall is perfect, but the experience of the spaces is so enjoyable that the audience perceives what it is hearing as acoustically ideal.

If Toyota's box looks back to the Musikvereinsaal, the democratic design of Gehry's interior—with no box seats or sky boxes and almost forty percent of the seats surrounding the orchestra—owes a debt to the Berlin Philharmonie. The architect says that he wanted "each listener to feel as if the orchestra were playing for him." The impression of being sensually wrapped in sound in this concert hall testifies to his success in this respect.

Richard B. Fisher Center
for the Performing Arts at Bard College

Annandale-on-Hudson, New York
Gehry Partners, 2003
110,000 square feet
$62 million

The Fisher Center could be called the poor man's Disney Hall. Only the principal facade of the much smaller upstate New York structure boasts the seductive stainless-steel forms that characterize the entirety of the Los Angeles facility. (The sides and rear of the Bard hall are plain vanilla—white plaster over concrete; load-bearing concrete and steel make up the building's structure.) And the Fisher's steel-and-raw-concrete lobby and corridors, having been cramped by budget-related cutbacks, lack

Exterior view.

the soaring quality of Disney Hall. The Annandale building has, however, the advantage of a beautiful, open meadow site in which its modest metal forms hover before a dense wood as gracefully as a giant pile of shimmering leaves. And the lobby compensates for its reduced depth with a four-story Piranesian web of metal trusses, struts, braces, and geothermal energy systems.

Designed about a decade after Gehry won the Disney Hall competition and several years before he started to work on the New World Center in Miami Beach, the Fisher Center incorporates aspects of both. But the Fisher hall is a multiuse proscenium theater, unlike the other two. There are important differences in program between the static, primarily single-use Disney Hall, the intermediary multiuse Fisher, and the extreme flexibility of the New World Center; in addition, the last two are teaching facilities.

At Bard the symmetrical, lyre-shaped, nine-hundred-seat auditorium, its smooth concrete walls and the fronts of its three balconies detailed in wood, is a classic theater—except for eight movable towers that can be rolled into place, together with a wooden proscenium whose opening can vary in size. The pale Douglas fir towers,

Auditorium,
without towers.

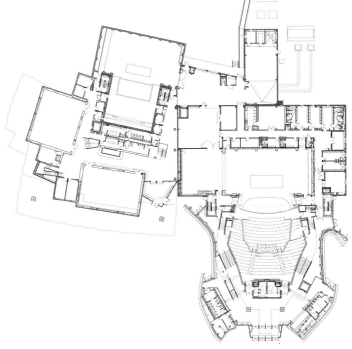

Lobby; plan.

twenty feet tall, are needed for the enclosing shell that transforms the theater from opera to concerts, with the musicians onstage rather than in the pit. According to Leon Botstein, conductor of the American Symphony Orchestra and president of the small college, which specializes in the visual and performing arts, the change is time-consuming (up to eighty man hours) and therefore expensive.[20] The center also contains a two-hundred-seat black-box theater, rehearsal studios, support spaces, and offices. While praising his theater's ability to serve a variety of art forms, Botstein concedes that, like Gehry, he feels a single configuration is best. Unlike the mediocre acoustics of older multiuse halls, Bard's are a resounding success, deemed excellent by James R. Oestreich of the *New York Times*.[21]

The bucolic Hudson Valley setting of the Fisher Center is unusual. So renowned is this picturesque landscape (depicted in the nineteenth century by such eminent American artists as Frederic Church and Thomas Cole) that preservationists opposed the original riverside site proposed for the Gehry structure. The center is now located more discreetly in its meadow setting; like Wagner's Festspielhaus (see Chapter 1), it is out of time and place. If not approached with the reverence intended by the German composer in the nineteenth century, the Fisher Center is at least free of the city's intrusions.

New World Center

Miami Beach, Florida
Gehry Partners, 2011
100,641 square feet
$154 million

Of Frank Gehry's three concert halls, the most recent, the New World Center in
Miami's South Beach, and the first, the Walt Disney Concert Hall in Los Angeles,
are at once the most similar and the most different. While Disney Hall remains
a landmark of the twenty-first century, nearly every guiding principle of that hall
is overturned in the 756-seat New World Center auditorium. Yet both the Los
Angeles and the Miami Beach halls are single rooms with no proscenium.

In contrast to the painful, sixteen-year birth of Disney Hall, the Miami project
offered Gehry the opportunity to work with his lifelong friend conductor and
composer Michael Tilson Thomas. Also a big difference between the two designs
was Gehry's resistance in Los Angeles to adjustable features in favor of what he
calls "a nice room in which nothing changes"[22] and his willingness in Miami Beach
to accommodate a dazzling array of transformations.

Tilson Thomas founded the New World Symphony in 1987 as an orchestral
academy for gifted graduates of major music conservatories. Attentive to the wishes
of the late Ted Arison, the founder of Carnival Cruise Lines, who spearheaded NWS's
$75 million endowment, Tilson Thomas chose Miami for the symphony's home. Two
parking lots offered a site for the three-part project: a new NWC building, a 557-car
garage, and the 2.5-acre Miami Beach Soundscape park, from which the public can
view free simulcasts of concerts projected on part of the center's facade.

About three-quarters of the new building is given over to a six-story atrium
housing student facilities, related offices, and cozy conversation areas. An aluminum frame system, attached to the building's steel frame, supports the atrium's
eighty-foot-high curtain wall on the east; a similar curtain wall encloses the atrium
to the west.

Floating within the all-white, six-story skylit atrium, which is traversed by a spiraling stair, are sculptural volumes that house individual and ensemble practice rooms
and a large multiple-use pavilion for orchestra rehearsals and small performances.
"They are built," says Gehry, "like a multistory village, like a city on a stage"[23]—modern
aediculae for a twenty-first-century version of the Teatro Olimpico's stage set. The
shaped rooms of the NWC atrium and the auditorium are enclosed within a rectilinear envelope that recalls another departure from this architect's usual curvilinear

exteriors: at the mixed-use, ten-story DZ Bank Building (2000) on Berlin's Pariser
Platz, his eye-catching forms were likewise confined to an interior atrium.

On the top floor is a music library, the patrons' lounge, and a private suite. A large
portion of the rooftop is enhanced by a garden designed by the landscape architect
Raymond Jungles, who selected the native south Florida trees and plants for which
he is known. It is open to patrons and concertgoers as well as to the fellows and staff.
Public and private areas are subtly divided, but views of the concave, convex, and
tilted forms within the atrium's private domain visually expand the concert hall
foyer. From the foyer, two low, narrow corridors snake into the center of the auditorium, entering it from the east and west near the front of the stage. Tiers of seats
rise steeply on all sides in an arena layout; where and how these seats are positioned
varies according to what is being performed and by whom.

Tilson Thomas's commitment to the visual aspect of presenting classical music
encouraged Gehry to devise fourteen different stage configurations within the
hall's trapezoidal container. Additionally, 247 seats can retract to allow a flat floor.
Four satellite platforms in different locations serve spatial music and didactic
demonstrations, such as presenting different musical genres by a single composer—
as for example during opening week, when the symphonic, piano solo, and

chamber work of Schubert were offered. Time-consuming re-sets of the main stage are thus avoided.

Concerts include theatrical, immersive lighting, specially commissioned videos, and contextual information (instead of program notes) projected onto the five huge, acoustic sail-like wall panels suspended below the ceiling. Of paramount importance to Tilson Thomas are connections to the Internet 2 network; in addition to making a concert in Miami available around the globe, this system allows New World members to take part in online projects and receive lessons from musicians elsewhere.

Internet technology has been used before, in one case, in earlier stages of development at the Judy and Arthur Zankel Hall in Manhattan (2003). However, NWC's acoustician, Yasuhisa Toyota, says, "The equipment is not new, but increasing the technological possibilities to this extent is new."[24] The concept has paid off. In addition to the hall's global connections, simultaneous projections on the sails surround the audience in a unique way and can, with the appropriate videos, create images that reinforce the music—as opposed to "movie music" that serves the visuals.

Exploded axonometric; atrium.

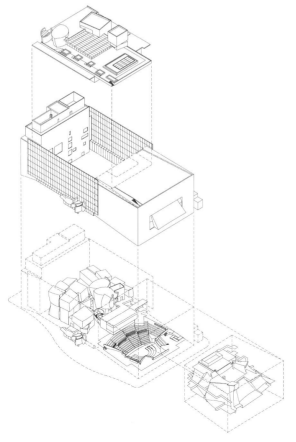

No seat in the auditorium is more than thirteen rows from the stage. The acoustic Auditorium.
sails swooping down the walls' upper halves visually lower the ceiling, adding to
the space's feeling of intimacy. Daylight is plentiful, entering through a skylight and
a generous horizontal window behind the stage, pointing up the stunning whiteness of the interior. Acoustical hardwood slats affixed diagonally on the partitions
between seating areas resemble garden trellises and, together with the Douglas fir
of seatbacks and arms and the blue-and-turquoise-patterned upholstery, evoke outdoor informality.

The concert hall is protected from unwanted sound by the massive concrete-
block exterior wall, separated by over three feet from the interior wall. Among the

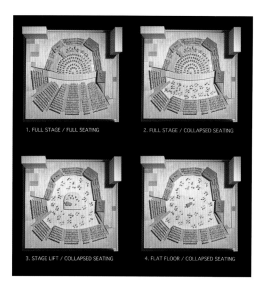

1. FULL STAGE / FULL SEATING   2. FULL STAGE / COLLAPSED SEATING

3. STAGE LIFT / COLLAPSED SEATING   4. FLAT FLOOR / COLLAPSED SEATING

Plan configurations.

auditorium's many acoustic features is the reflectiveness of both the stage's yellow Alaskan cedar and the ceiling baffles' plaster coating. The room's fifty-foot height (similar to Disney Hall's) and its boxlike shape are among the various elements that produce acoustics that are clear and enveloping no matter what size group is performing. An important and often overlooked aspect of acoustics is reported by the New World Symphony flute fellow Matthew Roitstein: "In contrast to our old Lincoln Theater hall [a retrofitted movie palace], now the musicians can hear themselves and each other."[25] Some complaints of excessive loudness will certainly be answered as the musicians adapt to their new home.

The painted-plaster exterior walls framing the atrium's clear-glass curtain wall inflect slightly, drawing attention to views of the interior, where even activities within some workrooms are visible. A wavy white canopy marks the entrance, beside which the sculptural box office, faced with plasma walls on both sides that announce programming, reaches out to the sidewalk.

During its opening week, close to a thousand people filled the 2.5-acre Soundscape park (designed by the Dutch firm West 8) to watch the Wallcast of each concert being played inside. Seated on collapsible chairs and reclining on blankets, people of all ages, many sipping wine and picnicking, were able to see close-ups of orchestra members and views of the ensemble, and to hear sound relayed via 167 speakers fixed into a grid of horizontal and vertical tubes that provides a close approximation of the music inside the hall (thanks to the electronic architecture devised by Fred Vogler of Sonitus).

The hall and its overall whiteness harmonize with the heart of South Beach's Art Deco redevelopment district, located a short block east on Lincoln Road. Gehry says he tried to capture in his design Tilson Thomas's vision of openness and accessibility: "It is a place that wants to be part of the community."[26]

Wallcasts are just one of the conductor's many ways of finding "a more spontaneous way to experience music than what is generally available, for example, the musical equivalent of an impulsive fifteen-minute museum visit."[27] These heady ideas and the heady concert hall stemming from them are off to a promising start.

Muziekgebouw aan't IJ/Bimhuis

Amsterdam, The Netherlands
3XN, 2004
170,608 square feet
$67 million (approximate)

This classically modern, low-cost music building with its spectacular views of the water is perfectly suited to the former dockside it occupies on the bank of the IJ River. Considered within the context of contemporary architectural acrobatics—exciting though they may be—the simplicity of the Amsterdam theater's glass-and-steel box, with some terra-cotta cladding at the lower level, fits well its shore site. This structure is the equivalent of comfort food at its best rather than gourmet dining.

3XN is a Danish firm founded in 1986 by principal Kim Herforth Nielsen. It began to work outside Denmark only in the late 1990s, notably on the Royal Danish Embassy in Berlin (1999). The architects' friendly, functional Scandinavian style appealed to the Muziekgebouw's director, Jan Wolff, who supported 3XN in the public tender to

Exterior view.

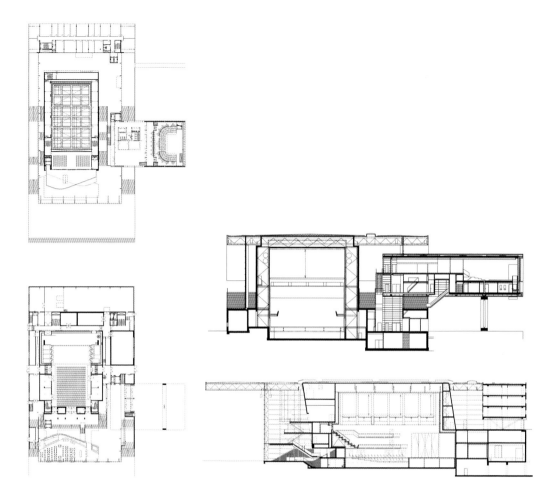

select an architect. The firm's extensive use of untreated tropical hardwood within the industrial-looking container makes for a relaxed, informal environment appropriate to the gritty history of the site, to its cutting-edge programming, and to the Bimhuis jazz club that is part of the complex.

Although younger (they were founded in the 1970s), and without the same aura as Amsterdam's famed Concertgebouw concert hall, both the Muziekgebouw for contemporary music and the Bimhuis for popular jazz and other improvisational genres are well-established institutions. They have functioned in separate locations since their beginnings. A simultaneous need for new premises provided a reason to share one of a series of structures slated for Amsterdam's eastern docks, a derelict area for which three master plans for renewal have been conceived since 1981.[28]

A prominent roof plinth (illuminated at night) gives the music facility a presence beside the much larger towers planned for the area. The facade facing back to the city is characterized by the assertive black-steel-clad box of the Bimhuis, which juts out beyond the Muziekgebouw's entrance. Running beside and emphasizing the bold protrusion is a lengthy (nearly two hundred feet long) bridge over flowing water, designed by Hans van Heeswijk, that provides an out-of-the-ordinary pathway to the upper lobby.

Lobby, with restaurant at lower right.

Under the bridge are a main, street-level entrance to the ticket booths, the primary concert hall doors, and a restaurant. Inside, a grand stairway connects this ground level with lobbies above and below, where rehearsal and dressing rooms are located. These expansive circulation spaces and the restaurant all enjoy natural light with views of the river to one side and the cityscape to the other. The indoor foyer functions also as a large plaza in which artistic and social events are held; in clement weather, these spill outside to the broad, wood-covered deck that extends the width of the building's north side. With an outdoor terrace at the top level and two wide outdoor stairways at either side of the structure—all in wood—the Muziekgebouw gives an importance to the exterior equal to various uses of the interior.

The 729-seat auditorium maintains the homespun feeling of the lobbies. Maple slats are affixed to the walls, and timber paneling and seat frames are a cool beige, with highlights of red provided by the upholstery. 3XN worked with the acoustical consultant Peutz to develop a space fully insulated from outside sound by a concrete box within a box and by two separate supporting roofs above the main auditorium. The auditorium itself is a model of flexibility: acoustically, it targets the clarity needed for twentieth-century music, but the reverberation can be adjusted from 1.5 to 3 seconds to accommodate other music; the hall is capable of assuming twelve

Bimhuis interior;
auditorium.

different arrangements, with each transformation taking no longer than an hour. Contributing to this mobility are ceiling panels that can be moved in height between approximately thirty-six and seventy-two feet and also mobile floor sections and seats. Colored lighting is available around the entire hall. Chinese-born Huang Ruo and Briton George Benjamin, two composers who have heard their pieces performed here, are lavish in their praise of the hall's acoustics. Huang notes that it is best for small ensembles.[29]

A priority for Huub van Riel, director of the Bimhuis, was to maintain an identity distinct from that of the contemporary music group. Retaining the same layout as its previous space, the one-hundred-seat jazz club occupies a long, low-ceilinged room within a discrete volume. A huge picture window opens the entire facade to views of Amsterdam. The black walls and restricted space (the interior was designed by De Vries Bouma) evoke jazz in contrast to the light-colored, airy concert hall for new music. The two venues have fulfilled the planners' wish to bring activity to this far end of the dockland, attracting 250,000 people from all over Europe in its first two and a half years.

Rem Koolhaas, who designed the second master plan for the area, emphasized the importance of keeping the profile compact and dense. Although this approach was for the most part retained, it is regrettable that two high-rises were built so near the Muziekgebouw that they detract from this handsome building.

Casa da Música

Porto, Portugal
OMA, 2005
247,000 square feet
$132.8 million

The odd polyhedron form of Rem Koolhaas's concert hall for Porto and its perverse positioning in the middle of a frenetic traffic circle are just what is needed to call attention to a facility that could redefine music in this region. As at OMA's Seattle Public Library, completed the year before, the unexpectedly irregular shape of the structure breaks with preconceived images of its function.

There is apparently reverse psychology at work here. The exterior is forbidding, its striated white-concrete surfaces glaring in the sun, but it stimulates curiosity and attracts the public. The Casa da Música's location, in a former trolley yard beside a

Exterior and plaza.

Stairway; found space.

small circular park that interrupts the broad Avenida da Boavista, isolates the facility within lanes of ongoing traffic. In its defense, the glazed facades of auxiliary spaces on the lower floors reveal their inner workings to pedestrians on the plaza, and the latter's wavy surface (atop a six-hundred-car garage), paved in rusty Jordanian travertine, has become a favorite hangout for skateboarders.

The seemingly impenetrable faceted monolith is entered through a grudging, twenty-odd-foot-tall slot slashed into the facades of two contiguous walls. Concertgoers climb to this entrance on a broad, bridgelike stairway—backlit at night, with balustrades of unframed glass—that feels more like a ship's gangplank than a grand access.

Any reservations about the exterior quickly succumb to the seduction of the interiors. The sliding glass door of the entrance opens to a minuscule foyer (and a wall of ticket windows) that is the starting point for a twisting, soaring, majestic staircase with steps finished in brushed aluminum. Pierced by dramatically angled trusses and lit by tall neon tubes—behind metal mesh or outlining the ceiling's cubelike forms—the stairs lead to the 1,250-seat Grand Auditorium and beyond. Other spaces are grouped at either side and on top of the auditorium. Among these are a red-stained 350-seat chamber music hall, a VIP room unexpectedly lined with traditional Portuguese *azulejo* tiles and furnished with antiques, a green cyber-music room, an education area, ten rehearsal rooms, and the usual restaurant (on an upper level) and bars, one a transparent glass box suspended within the building's stainless-steel structural members.

A walk through the Casa da Música affords views between the various levels (seven levels above ground, three below), between various rooms, and to the exterior (even to the Atlantic, visible from a rooftop amphitheater with a retractable skylight). Smaller, found spaces between the structure's diverse rooms and its envelope—a self-supporting, continuous folded plane—have stepped seating to encourage informal meetings and study groups.

Koolhaas describes Casa da Música as a solid mass from which the main auditorium and the chamber music hall are carved out. The former is constructed on a floating slab supported by acoustic isolation bearings. Updating the traditional shoebox concert hall, the architect has glazed the two shorter ends—one facing an old working-class area to the east, the other looking to the avenue at the west—and he has eliminated the proscenium. Because flat glass surfaces act against even diffusion of sound, OMA developed acoustically sensitive, corrugated and laminated glass for the windows. These are double depth at the sides and ends, with one aperture in the outer and one in the inner wall for sound isolation (from the exterior) and reflection (within). To prevent vibrations in the glass from entering the hall, the outer and

Tiled room.

inner windows are connected by a heavy steel frame that is vibration-isolated from the inner box.

A cushionlike object made of PVC with a soft surface occupies the place of a traditional large canopy over the stage; air pressure, height, and inclination can be changed mechanically to adjust acoustical resonance and reflectivity. A bold, embossed gold graphic enlarges the grain of the interior walls' plywood veneer, one of the architect's typical plays on cheap versus expensive materials. The floor in the auditorium is aluminum on concrete; the timber stage floor is painted an aluminum color and hovers over a layer of air. On the side walls of the stage, non-functioning replicas of two Baroque organs—made of plastic tubes, wood, and plastic molds—were required by the acoustical consultant, Renz L. C. J. van Luxemburg, when the real ones were dropped to reduce costs.

Despite the attention to acoustics, there has been criticism of the quality of sound in the hall. Composer Tod Machover, head of MIT's Media Lab, complains of dryness and a flutter echo, which, he says, are overcome only by drawing the knotted-nylon acoustical curtains designed by Petra Blaisse of InsideOutside.[30] Van Luxemburg is, in fact, working on acoustical adjustments: early 2011 saw the initiation of a remounting of the quadratic residue diffusers, which, by dispersing reflectivity more evenly, compensate for the lack of lateral reflections, which in larger halls would normally be provided by generous balconies. Wall and ceiling surfaces are also being treated.[31]

Renzo Piano's Niccolò Paganini Auditorium in Parma, Italy (2001), built within an abandoned sugar factory, has glazed facades at both ends; such departures from the usual auditorium enclosure now include Porto and several other concert halls,

Section; plan.

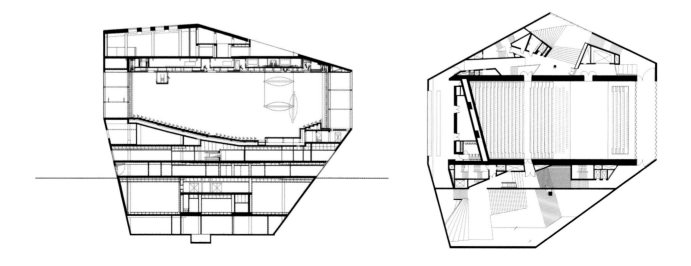

among them the Shalin Liu Performance Center (2010) in Rockport, Massachusetts (Epstein Joslin Architects), that offer the audience outlooks to the exterior. The feature has been welcomed by some as enriching and dismissed by others as distracting, but it ultimately reflects the architect's acknowledgment of the facility's context.

The slimness of the Porto hall, the steeply raked main seating with a few rows behind the stage, the warm gilt-decorated wood paneling, and the absence of a proscenium all contribute to an alluring feeling. Audience and performers are brought together in this room, which fulfills the concert hall ideal of a single, unified space. Once its acoustical glitches are worked out, the Casa da Música should provide a most pleasurable listening experience.

Danish Radio Concert Hall

Copenhagen, Denmark
Ateliers Jean Nouvel, 2009
210,000 square feet
$325 million (approximate)

Jean Nouvel has been to the performing arts what Renzo Piano is to the visual arts, a status now threatened by Zaha Hadid at Guangzhou and elsewhere. Beginning with the entirely new Opéra Nouvel (completed in 1993) that he constructed within the walls of Lyon's 1831 Opera House, and continuing with the Culture and Convention Centre in Lucerne (KKL; 2000), the Guthrie Theater in Minneapolis (2006), and the Copenhagen concert hall, with at least one more on the drawing board—for Paris's Ile Séguin development—Nouvel has designed and built more noteworthy theaters than any other architect (as Piano has built more noteworthy art museums). It is almost possible to credit Nouvel with the Copenhagen Opera House (Henning Larsen, 2006), so indebted is its design to KKL in the extensive cantilevered roof and the visibility from the exterior to the auditorium's outside wall.

The French architect's affinity for extravagant gestures is apparent in a number of his other buildings, notably the Paris Musée du Quai Branly (2006), in which anthropological displays are spotlit like stage sets within the darkened museum. Indeed, it is with a note of theatricality that Nouvel describes the auditorium of his Danish building as "a meteorite, fallen to earth in a desert."[32] The image is apt: a dome pokes above the roof of the concert hall and is covered with jutting, earth-colored "scales." Nouvel's presence in Denmark is in itself meteorlike: in the late 1990s, when plans for the competition for the Danish Radio Broadcasting Building, which contains the concert hall, were afoot, it had been more than two hundred years since a non-Danish architect (the Frenchman Nicolas-Henri Jardin) had worked on a major public project in Copenhagen.

The architect's "desert" is the Broadcasting Building's bleak site in a neighborhood just beginning to rise at the edge of the old city. This Ørestad North development, connected to Copenhagen by an elevated metro, is one of ten mixed-use areas planned by public and private initiative to replace obsolete industrial and shipping facilities. The massive project is not expected to be completed until the 2030s or 2040s.

In the absence of a context, to which Nouvel tends to relate his work, he designed a structure that posits its own urban presence within a small group of anonymous corporate buildings. The royal blue glass-fiber mesh that covers each side of the 592,000-square-foot glass cube is meant to evoke the background of a television

screen: as blank as an imageless television in daytime, the facades at night are animated by continuous projections of huge abstracted, music-related forms. A vast
glazed foyer, with additional video projections and commanding sights of the exterior,
is open to the public twenty-four hours a day and serves as a general gathering place
for those headed to the concert hall, restaurant, and bar.

This is architecture that is full of surprises: the exterior and interior video images;
the wrinkled, elephant-skin finish of the concrete foyer walls; the surreal effect
of people on six levels of steep escalators (supplementing elevators and staircases);
the glimpse of music studios below grade; the climactic auditorium. Smaller-scale
elements—galvanized-steel floors, brightly colored ceilings, and strident red ducts in
the office area—are also unexpected.

A poured-in-place concrete structure supports the 1,809-seat concert hall at the
upper part of the building. Its vineyard seating relates to the Berlin Philharmonie:
fourteen angular balconies of different sizes are cantilevered within a container that
is smaller and slightly more asymmetric than Berlin's. Emphasizing the asymmetry is
a rectangular balcony jutting from one side, which was designed for the queen of
Denmark and her family. The auditorium's upper portion alternates grooved walls of
birch-veneer board (also used on the ceiling) with undulating gypsum board painted
in autumnal colors; images were developed by the artist Alain Bony from Edvard
Munch's *Scream* and the Nordic sunset. The light-colored, watery pattern of the floor's

Belinga wood completes the decor. At a children's concert of light classics in December 2009, it was easy to perceive the rich sound not only of the orchestra onstage but of musicians playing from different areas of the auditorium, despite a lack of specific provision for such placement.

Technical equipment is isolated in a movable, eighty-five-ton sound-reflector canopy attached to the ceiling seventy-seven feet above the stage. The hall's acoustics (engineered by Yasuhisa Toyota) have been praised, albeit with calls for some adjustments. After conducting one rehearsal and one concert in the new hall shortly after its completion, Esa-Pekka Salonen described it as "a work in progress in every way."[33] Acoustically, the conductor finds the auditorium somewhat dry. He is also concerned about the concert hall's isolation from the city: "Now it is like an ecosphere, but you have to be part of something to thrive." Conductor and pianist Daniel Barenboim complained that the orchestra has trouble hearing itself.[34]

Three small additional music spaces are located one level below grade. Studio 2, seating 540 and adorned with enlarged photographs of famous music professionals, is used for rehearsals and concerts. The black-and-white-lacquer decor of Studio 3 immediately suggests jazz, for which the 170-seat room is intended. Red aluminum and felt panels pivot for acoustical flexibility in the 180-seat Studio 4, used for recording and rehearsal; Nouvel acknowledges that its walls of adjustable panels,

Plan; foyer.

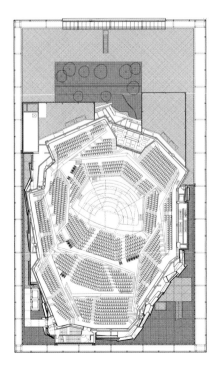

which vary sound absorption and reflectivity, were inspired by IRCAM's experimental space (see Chapter 1).

Auditorium.

Regrettably, the roughshod lobby of the Danish Broadcasting Building does not live up to the luxurious sensuality of all four music spaces, especially that of the main auditorium. Contrasting scale and materials are standard architectural devices, but in Copenhagen Nouvel has taken these juxtapositions too far. The lobby's cavernous size and excessive use of plywood, including flimsy-looking furnishings designed to look like flight crates for musical instruments, produce a feeling of impermanence, as if the area were still under construction.

This impression is not helped by the lighting. After nightfall, when most theaters come to life, the immense lobby, with its dark concrete floor, is characterized by a gloominess not sufficiently relieved by light or video projections. Perhaps lighting adjustments will be able to approximate the cheerful sunlight that adds so much to this facility during the day.

Because Nouvel always responds to a specific situation rather than to any universal norm, it is impossible to ascribe a particular style to his work.[35] This unpredictability is a key to his success with theaters, in which surprise is, after all, an essential ingredient.

Koerner Concert Hall,
TELUS Centre for Performance and Learning,
Royal Conservatory of Music

Toronto, Canada
Kuwabara Payne McKenna Blumberg Architects, 2009
TELUS Centre: 190,000 square feet
Koerner Concert Hall: 30,000 square feet
$83 million

When performance space is added to older halls, it is almost invariably inserted within an existing building, as in the case of the Judy and Arthur Zankel Hall beneath Carnegie Hall and, more recently, the rooms beneath the Musikverein in Vienna. The 1,135-seat Koerner Concert Hall, by contrast, is part of a freestanding building, the TELUS Centre, that wraps around two sides of Toronto's 125-year-old Royal Conservatory of Music (RCM). The astute use of a diminutive site for the center, which now blends new and old, and the resulting elegant, low-key concert hall augur favorably for other such expansions.

Marianne McKenna, a founding member of Kuwabara Payne McKenna Blumberg, has had a long relationship with the RCM, Canada's equivalent of the Juilliard School in the United States. In 1990, McKenna produced the award-winning master

Site plan showing original conservatory (top) and addition.

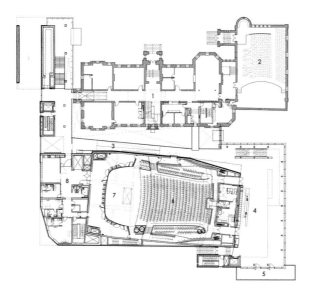

Existing building (left) and new Bloor Street entrance; atrium.

plan that included restoration of the nineteenth-century buildings, and since then she has steered a number of renovations within the complex.

The TELUS Centre is entered through a narrow three-story modern structure that reaches out beside the original institution, a handsome red-brick Victorian building, toward Bloor Street, Toronto's premiere shopping artery. The attractive entranceway consists of a glass-enclosed ground level topped by a slate-sheathed box containing an airy rehearsal room—the largest of several studios—with continuous floor-to-ceiling windows in one corner. Passing the box office and ascending a limestone stair, visitors find themselves on a bridge that crosses a skylit three-story atrium—trapezoidal in plan—between the historic and the contemporary structures.

This dynamic area, overlooked by small balconies and a mezzanine, ties together the disparate spaces. Like other expansions that enclose former facades—the Metropolitan Museum of Art's European Sculpture Court in Manhattan is a similar example—this one transforms the conservatory's aged, polychromatic wall into a colorful backdrop, in this case for the entranceway to the concert hall from the mezzanine and, below it, for a lounge and café. Animated by views of students behind the glass walls of surrounding corridors, the atrium is a pleasant place to be in and also functions efficiently, bringing the public deep into the site, where the concert hall is located.

Outstanding in the Koerner's tall, slim shoebox are the twisted ribbons of laminated white oak that screen the thirty-nine-foot-high ceiling and layered front of the room. McKenna says, "The 'veil' was my original architectural inspiration for a dramatic way to conceal both the fixed acoustic canopy and the rigging above it. As

the idea evolved, the oak strips, which twist under the canopy, allowed us to dispense with the acrylic acoustic rings that disperse the sound from the stage."[36] The technical canopy resides in the sixteen-foot-high space that separates the ribbons from the concrete ceiling. The inflected shapes of the ribbons and the gaps between them are among a number of features that are acoustically effective and also allow lighting and speakers to be interspersed. Behind the stage, the veil scatters sound in back of the musicians the way organs do in other halls. Robert Essert, the project's acoustical consultant, calls the effect "a river of flowing timber."[37] It also brings to mind the flow of music. The wavy theme continues with textured light brown plaster panels on the side walls and two slightly curved balconies fronted with convex oak boards.

The narrow balconies surround the hall, positioning ten percent of the audience behind and to the sides of the stage, which is without a proscenium. An adjustable system of riser platforms fits into the stage, lifting the musicians at the back when desired. Altogether, the sensuous, rounded shapes of this interior coupled with the light color of the nearly ubiquitous timber make it captivating.

The Koerner's acoustics have worked well whatever the scale of the music: from a piano solo to a large-scale choral piece, the sound has clarity and warmth.[38] Writing in the *Globe and Mail*, the music critic Lisa Rochon described the sound as "stunning: urgent, velvet, intimate."[39] As in most new concert halls, a set of sound-absorbing curtains can be used to tone down the reverberant acoustics and accommodate percussion or amplified concerts. To keep out noise from the subway under the site, from the chillers in the hockey arena next door, and from the teaching studios and mechanical equipment rooms, the hall was built as a separate concrete structure floating on thick rubber pads. (A similar system isolates the rehearsal hall acoustically.)

At the addition's easternmost corner, diagonally across the site from the Bloor Street entrance, the three-level glass curtain wall of the lobby links the auditorium with the city. Running alongside the building is Philosophers' Walk, a tree-lined pedestrian path to the University of Toronto, with which the RCM was once affiliated. Almost immediately opposite is the Royal Ontario Museum's sculptural rooftop VIP room, designed by Daniel Libeskind as part of his recent Michael Lee-Chin Crystal addition (2007). Besides Libeskind, Frank Gehry (Art Gallery of Ontario, 2008), Will Alsop (Ontario College of Art & Design, 2004), and Jack Diamond (Four Seasons Centre for the Performing Arts, 2006) are among the growing list of well-known architects who have contributed to Toronto's cultural facilities; the TELUS Centre adds McKenna to this roster.[40]

Auditorium.

Haus für Musik und Musiktheater (MUMUTH)

Graz, Austria
UNStudio, 2009
66,736 square feet
$23 million (plus $7–10 million for technology)

The historic onion-shaped towers of Austria's second-largest city are deceptive: Graz has been remarkably daring in its commitment to progressive architecture. Peter Cook and Colin Fournier's nozzled, biomorphic Kunsthaus Graz started the trend in 2003, when the city was designated that year's European Capital of Culture. It was soon followed by Vito Acconci's complex Murinsel (Island in the Mur), a mid-river glass-and-steel pleasure pavilion that includes an open-air theater, café, and playground.

UNStudio's Music House (MUMUTH) is the latest exemplar. The local art university (Kunstuniversität Graz), which is committed to contemporary music and science, decided in 1998 to construct a theater in which its 2,100 international students could have a foretaste of public performance, along with rehearsal rooms, workshops, and a lounge. The winning competition entry, by UNStudio founders Ben van Berkel and Caroline Bos, included a remarkable support structure that determined the shape of the building and the column-free spaces within: three interlocking spirals of steel pipes, stabilized by self-compacting concrete in the steel, frame walls, and floor slabs. When this design proved too expensive to build, a greatly simplified version was adopted. Two of the spirals were discarded; the third was reduced in size and its exterior expression eliminated. Furthermore, the spiral was transformed from steel to concrete, although steel was retained for the exterior.

Reflecting the facility's private and public functions, its glass-and-steel exterior is masked in daytime by a fine stainless-steel mesh attached to gently curved steel frames. Because MUMUTH is an academic institution, it is a mysterious, anonymous presence during the day, its private educational and administrative spaces reserved for students and staff. Only at night, when interior lighting reveals parts of the interior, does the presence of a public concert hall become apparent. Two separate entrances—public to the south and student to the west—facilitate the various uses.

Adopting mid-1990s parlance to explain the compromise of the original design, the architects use the phrase "blob-to-box model."[41] The movement-based volumes of the foyer and public circulation at the south are the blob; the unit-based volume of the theater at the north is the box. The two volumes appear unresolved in plan, but then a spiral joins blob and box, organizing the whole. Based on a Möbius strip, the spiral is a form that van Berkel and Bos began to explore in 1993 for their own home

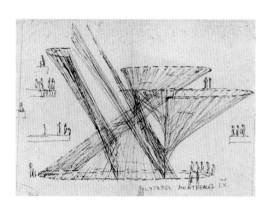

in the Netherlands. They liken the coil structure to serial technique in contemporary music: both share the ability to absorb and regulate intervals, interruptions, changes of direction, and leaps of scale without losing continuity. Indeed, although it is related to different musical principles, the helix-like form bears a striking resemblance to the composer-architect Iannis Xenakis's 1960s Polytopes, his model for electronic light and sound projection. Van Berkel's convincing parallel between architecture and music was the determining factor in the competition for the building.

The public is drawn from the south entrance, at ground level, toward the main stair leading to the music theater on the first floor by the expressionistic "twist" that connects the facility's three levels. This elegant connection makes a powerful statement, which is unfortunately weakened by the spindly red-carpeted, metal-encased stairway that it becomes as it leads from the theater level to the skylit floor above.

Compared to the structural tour de force of the blob, the sober 450-seat theater might appear tame. Its technology makes it anything but. The auditorium is basically a black box, its dark eggplant walls discreetly articulated with a low-relief version of the silkscreen pattern, based on musical instruments, that is printed on the building's glass curtain wall. However, 108 floor modules—measuring about six and a half feet in length and three feet in width—are designed to rise hydraulically to almost eleven feet (higher at the back). Only six technicians are required to control such changes, and Georg Schulz, rector of the university, says that the room is a stage that is "reconfigured constantly" for events other than classical concerts.[42]

Delay of the building's construction until 2006 allowed the installation of the most up-to-date electronic architecture (the Constellation System, similar to that used at the New World Center's Soundscape park). The hall has a reverberation time of just over one second (consultants were the local concerns ZT Gerhard Tomberger and Pro Acoustics); with the system, the reverberation time is adjustable to over two seconds, so the room can be sonically altered to suit a range of performances from the spoken

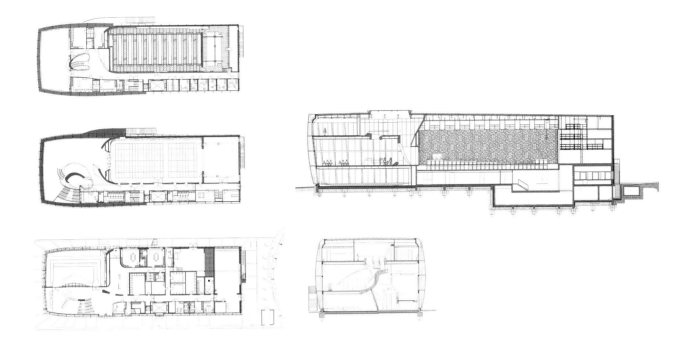

word to chamber music, opera, and jazz (a large orchestra is the exception). Additionally, it is used to replicate the different kinds of spaces for which the music was composed. For early religious choral music, for instance, the reverberation can be set electronically to re-create the acoustics of a large church.

For a presentation of Bach's *St. John Passion*, the floor modules were raised in a double-eight design by van Berkel (echoing the Möbius strip), which stretched throughout the room and placed the choir within the audience. The conductor and the orchestra were at one side, away from the raised area. Since the hall's completion, rearrangement of the seating and implementation of the electronic architecture system have enjoyed equal usage. The performance of music thus joins architecture as another area in which this city dares to experiment.

Glass Hall, Musikverein

Vienna, Austria
Wilhelm Holzbauer, 2004
2,476 square feet
Cost undisclosed

Wood, though hard and reflective, implies a warm sound, like that of a musical instrument, and is often articulated to diffuse sound reflections. In contrast, glass is a harder, smoother surface—both visually and acoustically—and accentuates bright, even brittle, sounds. This material has been used for music facilities only rarely, perhaps because of this association. And yet a recently designed small room with glazed walls is successful in every respect.

The Glass Hall, with 380 seats, is the largest of four small insertions—each in a different material—two levels below grade at the Musikverein in Vienna. Wilhelm Holzbauer, one of Austria's most prominent architects, was chosen for the job. With

Interior view.

his project partner at the time, Dieter Irresberger, he collaborated on the 1,500-seat Opera House in Amsterdam (1988) and the 2,650-seat Festspielhaus in Baden-Baden (1998). For the Glass Hall, they worked with the acoustician Karl Bernd Quiring.

The program specified a rehearsal room for the Vienna Philharmonic that would replicate the acoustical conditions of its principal Golden Hall with the audience in place (see Chapter 1). The architects built a stage with the same dimensions and the same four adjustable sections as the Golden Hall's. Mounted on a metal grille that stands a little more than three feet from the wall behind the stage are fixed and movable, flat and curved, glazed, rectangular glass panels that feature on their reverse sides a layer of glowing gold leaf. The space between the grille and the wall accommodates the machinery needed to adjust the panels for sound absorption. The other three walls, which support balconies (a one-row balcony facing the stage and two-row balconies at either end), are faced in matte, gilt-colored glass; ceiling panels are white glass; the floor is oak. The balcony's glazed balustrade panels are sparsely decorated with a digitized pattern, opaque at the base and transparent at the upper portion.

Holzbauer, a confirmed postmodernist, chose to work with glass so that his interior would look completely different from the historic one. While the Glass and Golden Halls do not relate visually, Holzbauer's use of gilt and the patterned balustrades evoke the Vienna of 1900. And access to the addition benefits from careful detailing characteristic of the earlier, Secessionist era: an elegant little elevator and a renovated stairway descend to a wide, brightly lit corridor, off which the Glass Hall and the other new rooms open.

A student recital by two small ensembles in December 2010 offered crisp, bright sound. The hall is acoustically extremely lively; audience sounds projected as much as those from the stage. Although big orchestral and choral pieces have been rehearsed satisfactorily in the hall, it is used more frequently for concerts of small ensembles, chamber music, and solo recitals.

Crystal Room

Waidhofen an der Ybbs, Austria
Hans Hollein, 2009
2,691 square feet
$2.5 million (approximate)

Hans Hollein is better known outside his native Austria than Wilhelm Holzbauer; his Städtisches Museum Abteiberg in Mönchengladbach, Germany (1982), was a major step in a postmodern rethinking of museum architecture, and his exhibition design for the inauguration of the Cooper-Hewitt Museum in Manhattan (1976) received considerable attention. Hollein has produced several handsome projects for concert hall competitions, all of which have remained unbuilt. Among them was his entry for the Disney Hall competition, a finalist. For the Crystal Room, he adopted a solution completely different from any of his previous ones.

The medieval Rothschild Castle in the quaint town of Waidhofen an der Ybbs, a two-hour drive west of Vienna, has been used in various ways by a succession of proprietors. Since 2003 it has been owned and controlled by the town government, which recently opened the building to the public. Hollein's renovation has made most of the space into a conventional history museum. Less predictable is the unusual concert hall within the upper floors, which is used for annual music

Rothschild Castle,
Waidhofen an der Ybbs,
medieval era.

festivals and year-long programs of classical and jazz concerts. The architect also crowned the castle's main tower with one of his favored rooftop gestures, in this case a twenty-foot-tall glass box that is illuminated at night.

Like Holzbauer at the Musikverein, Hollein selected glass as the primary material for the auditorium. He worked with the acoustician Karlheinz Müller of Müller-BBM to design the faceted, translucent, and laminated glass panels of various sizes and shapes that stretch across the walls and ceiling of the double-height, two-hundred-seat room. Joints are sealed vertically but not horizontally so that gaps between the tilted glass plates on the ceiling offer sound absorption and ventilation. Additional absorption is provided on the back wall by aluminum panels in front of a layer of mineral wool. Spotlights and loudspeakers are embedded between, and affixed to, the ceiling panels.

Existing windows behind the translucent walls allow sunlight to filter through, emphasizing the room's predominant whiteness. After dark the effect is sustained by electric light. Blue fluorescent uplights are concealed at floor level. Seating radiates around three sides of an off-center, elevated, semicircular stage.

Herbert Schuch, an accomplished young pianist, was at first surprised by the unusual placement of his instrument at one side of the room. Once he had adjusted, Schuch found the hall "magnificent: it was like being inside a crystal."[43] He continues, "I was energized by the light and by being very near the public." The most succinct endorsement comes from Markus Hinterhäuser, artistic director in 2011 of the Salzburg Festival, who describes the hall as an "architectural gem that is without doubt faithful to the message of music."[44]

JS Bach Chamber Music Hall

Zaha Hadid Architects, 2009
3,900 square feet (approximate)
Cost undisclosed

The Iraqi-born British architect Zaha Hadid came to public awareness with her project for the Peak Club in Hong Kong (1983). Like her other designs, it is characterized by such startling forms that many observers questioned the possibility of building them. Her Vitra Fire Station in Weil am Rhein, Germany (1994), silenced doubts and marked the beginning of a career that has produced major works worldwide, including the Rosenthal Center for Contemporary Art in Cincinnati (2003), the BMW Central Building in Leipzig (2005), and the Guangzhou Opera House (2010). The practice Hadid established in London in 1980 now includes over 350 people.

Commissioned as a temporary installation within a black-box gallery at the Manchester Art Gallery (for England's 2009 Manchester International Festival) and used for the same purpose at the WGF Gashouder, a retired gasworks in Amsterdam (for the 2010 Holland Festival), Hadid's JS Bach Chamber Music Hall is a broad white Lycra ribbon stretched over a steel frame. It envelops approximately 190 listeners in sensuous twists, spiraling around and above them, evoking the billowing scarves of modern dancers and making a womblike enclosure for the performers and audience.

Hadid's design expands the tradition of twentieth-century essays in small spaces for which music was composed, such as Le Corbusier and Xenakis's freestanding Philips Pavilion in Brussels and Renzo Piano's infinitely repeatable fixed module for Luigi Nono's *Prometeo*, first erected within the Church of San Lorenzo in Venice (1984; see Chapter 1). Hadid's temporary structure is different: with each deployment within an existing space it changes in both width and orientation. Despite its name, the pavilion may be used for different kinds of classical music, and there are also plans for poetry readings and storytelling.

The acoustician Mark Howarth (of Sandy Brown Associates) emphasizes the care that was devoted to the choice of Lycra—in this case, a lightweight synthetic that would not reflect sound from the installation's curves but at the same time would avoid being overly absorbent. Around the stage, specially shaped acrylic reflector panels are hidden within the curves. While these considerations play a part in the pavilion's acoustical success, that success also depends on the space in which it is installed.

A circular neo-Renaissance brick structure, the voluminous Gashouder (1903) has a high roof held up by radial beams, with no need for supporting columns. Within this space, the pavilion worked phenomenally well for a recital of complex

contemporary Dutch compositions for piano, which included both natural acoustic and electronic sound, and in one case a disassembled piano.

Hadid's minimal ribbon retains the notion of a classical stage and even of a corridor at the back. Within the small hall, every member of the audience enjoys a direct relationship with the performer. So effective is the visual impact of Hadid's pavilion that listeners should be delighted by whatever they hear within it.

Installation at WGF Gashouder, Amsterdam, 2010.

HISTORICALLY, NO OPERA HOUSE INTERIOR HAS BEEN AS TRANSFORMATIVE AS Scharoun's Berlin Philharmonie has proven for concert halls: Wagner's equally innovative Festspielhaus is acoustically satisfactory for only a small selection of works, and its imitators are limited. Conversely, the Berlin hall's vineyard design accommodates a broad musical range and has been widely adopted right up to the present and into the future.

With the introduction of a less formal, closer relationship between public and performers, Scharoun's design opened the way to a more intimate listening experience—a preferred condition of contemporary theaters. Furthermore, the vineyard arrangement successfully challenged the persistent acoustical taboo of asymmetry. Recent research holds that asymmetry can actually benefit acoustical quality by adding richness of texture to the sound. Listeners in the center of a symmetrical room (shoebox or horseshoe) hear the same sound with both ears. Those in an asymmetrical room perceive an additional sense of spaciousness, a discovery that gives architects unprecedented freedom.[45]

Opera is the most conservative genre within the already conservative culture of classical music. Worldwide programming allots a mere fifteen percent of the repertory to works more recent than the nineteenth century.[46] But even the most traditional repertoires can be enhanced by inventive staging and design. The point is proven with new technical capability for concerts in Gehry's New World Center, which integrate visual and audio offerings. The novel performance areas in Koolhaas's project for the Taipei Performing Arts Center (2015) has similar potential for opera; it could be the beginning of a critical mass of music theaters that are adaptable to individual works, facilitating a relationship for shared productions (see Chapter 5).

From the early twentieth century on, artists have encouraged viewers' active involvement with what they are looking at. Currently, viewer participation has become an intrinsic goal of museum display, with performance art one of its clearest manifestations. In 2010 New York City's Solomon R. Guggenheim Museum and Museum of Modern Art both featured performances in which the public participated, actually becoming an element of the artwork. Museumgoers answered questions posed by a series of guides as they ascended the Guggenheim's ramp in Tino Sehgal's *This Progress*. And as part of Marina Abramović's *The Artist Is Present*, individuals sat opposite, and engaged in eye contact with, Abramović in MoMA's atrium in full view of other visitors. Glenn Lowry, director of MoMA, called the Abramović exhibition the "culmination of a decade-long effort to change the museum's character, to turn it 'interactive,' a place where people come to see, but also to be seen; not just to look at art but to participate in it."[47]

A corresponding sense of audience involvement is now expected for musical performances, something that is encouraged by the user-friendliness of recent music venues. Each person who visits these spaces, while enjoying his or her personal impressions, consciously belongs to what the philosopher and critic Walter Benjamin (1892–1940) presented in his thinking about the rise of mass culture as a "simultaneous collective experience" (possible for architecture at all times, for the epic poem in the past, and for the movies at the time he wrote).[48] Thanks in part to seating arrangements that allow listeners to look at one another instead of at the backs of heads, a feeling of community is established—to a certain degree. The soprano Renée Fleming, for one, complains that it is precisely the back of her head that audiences seated behind the stage see, a situation she feels is unfair.[49]

In addition to designs that promote visual intimacy, these theaters benefit from acoustical engineering that, apart from ensuring excellent sound, insulates them from unwanted noise and produces an aural intimacy. Without the interference of subway rumblings, wailings of fire engines and ambulances, or climate-control mechanics, listeners can immerse themselves in the immediacy of their experience.

There is perhaps a lesson to be learned from the success of the participatory Sehgal and Abramović performances, which attracted, respectively, more than one hundred thousand visitors at the Guggenheim in one and a half months and over half a million at MoMA in two and a half months.[50] Certainly, the way these new sites for sound look influences the way audiences hear, but they offer more than spectacular architecture. Buildings for music that encourage their own kind of participation, comparable to that of performance art, may find this to be a key to their success.

4 China

"Building Big"

I N THE PEOPLE'S REPUBLIC, THE PROCESS OF DESIGNING AND building new spaces for music and, to a lesser degree, the experience of attending an opera or concert are distinct from such undertakings in the West. A city's decision to begin construction of a cultural complex, what is expected from it, and how an architect is chosen differ from the way these issues are handled in the United States and Europe. All aspects of such projects, like everything in China, are controlled by the government, be it the city, provincial, or national administration. In the absence, for the most part, of what is considered noteworthy or enduring architecture (the average lifespan of new commercial construction in China is thirty years; in the West it is open-ended),[1] these buildings must be evaluated on their sociopolitical role as much as on their design.

From the time that a clavichord was introduced to the Ming court by the renowned Jesuit priest Matteo Ricci in the early 1600s, Western music has had a rocky history in China. The country's internal strife and external confrontations have affected regard for everything foreign, including the arts. No time was more difficult in this respect than the twentieth century, when political upheavals culminated in the Cultural Revolution (1966–76), which destroyed scores and recordings; ravaged Western-style orchestras; closed conservatories; and "re-educated," wounded, and even killed musicians. The ban affected Western classical music along with millenniums of traditional music from the Imperial era; only propagandistic, Soviet-style compositions were tolerated. The same political events that stymied the country's musical development also distanced China from twentieth-century architecture. The country was thus left with a need to catch up in both spheres.

Following about-faces in China's government policies, attitudes toward culture started to change in the late 1970s, and by the early 1990s, with the country opening up to foreign investment and the beginning of its economic boom, music-related projects came into being. There are two interrelated reasons for the proliferation of buildings for concerts and opera in China: national and global. First is an intercity cultural competitiveness. Despite fierce resistance to plans for a new opera house in Shanghai, once it was completed, Beijing and other cities rushed to match the eastern metropolis with their own prominent facilities. Second, the central government intends to make China a leader in every area, including the arts, and thus is strongly supportive of new buildings for cultural events (continuing the Stalinist fondness for cultural palaces).

Since 1990, China has seen the greatest period of urban growth in human history. The People's Republic has achieved in one generation what it took the United States a century to accomplish, constructing in this time more infrastructure, more high-rises, and more of every other building type than, arguably, the rest of the world.[2] Besides an array of formidable skyscrapers—by 2014 the Shanghai Tower, at over two thousand feet, will be among the tallest structures anywhere—one of the most distinctive aspects of this surge is new architecture for the performance of music: the so-called Grand Theaters. (They are called *dajuyuan* in pinyin, the official system of transcribing Chinese characters into the Roman alphabet. The term has been inconsistently translated and inconsistently applied to various structures.) Like other new cultural buildings worldwide, Grand Theaters are favorite high-profile landmarks in otherwise undistinguished neighborhoods: in China, for the anonymous large-scale New Towns rising at a breakneck rate there.

Most Grand Theaters consist of three auditoriums: a relatively large opera house, a smaller concert hall, and a still smaller flexible space for theater. (The opera house is not, however, what that name implies to Westerners; rather it is what is generally termed a multiuse theater.) The models are Lincoln Center in Manhattan and the Kennedy Center in Washington, D.C. The Chinese government has sent many young administrators to the latter's fellowship program, and Lincoln Center agreed in early 2011 to serve as a paid advisor to the developers of a proposed performing arts center in Tianjin.[3]

As is the case for all construction of public buildings in the People's Republic, music theaters are subject to competitions for the selection of an architect. But China can carry to extremes the shortcomings of competition practices in the West. Foreign architects are often surprised to realize that they are the only non-Chinese presence on juries described as multinational, and a jury's choice counts merely as a suggestion to the relevant mayor. The government appoints every mayor, an all-powerful official who frequently overrules competition juries with his own preference for an alternative competitor, or even for an architect not included in the competition. So in the same way that important local political appointees are designated by a central bureaucracy, decisions about the design and construction of Grand Theaters are handed down from the highest local or regional authority.

Government decisions can be capricious, having political rather than knowledge-based motivations. In Guangzhou the local Design Institute—one of several thousand state-owned regional offices that replaced private architectural practices in China from the 1950s to 1976—won three consecutive competitions for the opera house before government officials ordered yet a fourth. This time, Zaha Hadid was awarded the commission. Once approved, work was repeatedly interrupted by struggles for control between the city and provincial authorities.

There are exceptions to such breaches of standard practices for competitions, one example being those held in Beijing for the Olympic projects. However, the Chinese have no compunction about copying losing entries and bypassing the originating architect; protection of intellectual property rights has generally been difficult to enforce.

Many plum commissions in China, including those for theaters, have gone to French nationals, and it is interesting to trace the reasons why. Following French president Jacques Chirac's 1997 trip to Shanghai, where he was greatly impressed by development in the Pudong district, he inaugurated the Franco-Chinese Presidential Program (1998–2001), in which fifty Chinese architecture students were given scholarships for a year's study in a French school with an internship in a French architectural firm. The program was so successful in creating a Franco-Chinese network of architects that, since its completion, it has been reinstated several times.[4]

Even before the Presidential Program was launched, the French Ministry of Culture and Communications supported French Architects Overseas (AFEX) to promote French architecture and related fields worldwide. It would be naive in this context not also to take into consideration the enormous trade deals between China and France, worth a reported $40 billion in 2010 alone.[5]

In the case of competitions for music facilities, there are traits characteristic of the People's Republic. Acousticians present together with the architect's schematic and development designs. For all building types, more detail is required than is usual for competitions in the West (and the fees are lower). Foreign architects accept these conditions because the lower cost of building in China (approximately one tenth that of the United States) along with the country's boundless ambitions and generous budgets affords precious opportunities to make great architecture.[6] Ole Scheeren, until early 2010 the director of OMA's Beijing office, says of the firm's spectacular China Central Television (CCTV) Tower there, "This building could never have happened anywhere else."[7] Steven Holl echoes this opinion in discussing his 1.3-million-square-foot Horizontal Skyscraper in Shenzhen.[8]

In the West, all designs are subject to local structural building codes, which often require modification; further revisions are carried out by some clients for non-code-related reasons. In China, the actual work of foreign firms must be executed in concert with a regional Design Institute in a far more collaborative arrangement than that typically required by the architectural licensing boards in the United States or elsewhere.

Until the twenty-first century, grandiosity rather than acoustics was the main preoccupation for China's new cultural centers. East Asia generally follows the Western tradition of show-stopping exteriors for opera houses. (The Sydney Opera House is often invoked as the model for a world-renowned exterior that attracts visitors to the city.) But when it comes to interiors, the attitude of Chinese clients toward sound differs from that of their Western and Japanese counterparts, for whom natural acoustics determine the success or failure of a venue.

Traditional opera houses have seats that surround the stage in a horseshoe configuration, and the orchestra pit is partially tucked under the stage to favor the singers' outward projection, without amplification. Chinese opera houses usually acknowledge this physical model (although the pit is occasionally completely uncovered), but they reverse the priorities of most Western houses, which are designed first and foremost for eighteenth- and nineteenth-century opera, and only secondarily for dance and acoustically amplified Broadway-type shows.

In addition to the customary horseshoe for opera houses (Hadid's discreetly asymmetrical configuration for Guangzhou is a rare departure from this practice), most Chinese concert halls have shoebox interiors with some variations on the vineyard (Xinghai Concert Hall, Shanghai Oriental Art Center, Shenzhen Concert Hall). Chinese halls tend to be a little wider than is the norm in the West—posing a challenge for acoustics—since walls are pushed back to reveal the supporting columns (also prominent in the lobbies) favored by the Chinese. Steeply raked, often continental-style seating is common, and a VIP area in every Chinese theater corresponds to the historic royal box in European theaters. Glass-fiber-reinforced gypsum, which has acoustical properties similar to plaster, is used for most auditorium walls. Air is circulated from the floor. Dressing rooms—usually with windows—and other support areas are generous. Handicapped access varies from theater to theater. Most of the opera houses, and even some of the concert halls, feel formal, without the intimacy that Western theaters now try to achieve.

Despite the prevalence of foreign architects for most Grand Theaters, local design customs—including the use of feng shui and geomancy, which call for curves to deter bad spirits that travel in straight lines—can give them a distinctly Chinese flavor. Elaborate floral decorations in the lobbies and around the stages are particularly abundant during holiday periods, when colorful fabrics and ribbons also dress up columns, balustrades, and balcony fronts. Posters for coming attractions, which as often as not conflict aesthetically with their context, are displayed year-round.

One of the outcomes of the Cultural Revolution was the insistence by Jiang Qing (Madame Mao) that the centuries-old, small-scale Peking opera be replaced by revolutionary opera modeled on the kind of propagandistic cultural events that were staged by the Soviets following the establishment of their regime in Russia. *The East Is Red* (1964) is typical of these Chinese music and dance epics, which call for casts of hundreds. The format has remained popular, as witnessed by the 2009 National

Center for the Performing Arts (NCPA) production to celebrate the sixtieth anniversary of the founding of New China, *Road to Revival*. More than two hundred dancers, three hundred soldiers, and two choruses of nearly one hundred people participated. The singers stood on a thirty-foot staircase for over two hours; sets changed every two minutes. Music was a karaoke-style pre-recorded accompaniment.

With this kind of extravaganza, and with Western operas like *Turandot* and *Aida* lending themselves to similarly elaborate productions, the stages, back stages, fly towers, and orchestra pits are relatively large and technically advanced. As for acoustics, these spaces effectively fulfill a multiuse function, thanks to variable sound-absorptive banners that can be deployed when necessary. Where the halls are less satisfactory is for Western-style orchestras playing either in the pit or on stage with the assistance of an acoustical shell. For these events, sound tends to be lifeless.[9]

Possibly influenced by the pervasiveness of public loudspeakers before and especially during the Cultural Revolution,[10] almost everything, occasionally even classical Western music, is amplified (as with pianist Lang Lang's flashy concerts in stadiums and concert halls throughout the country). Furthermore, visiting performers of popular music find it less expensive to bring in their own equipment than to rent the theater's amplification systems. Since the imported loudspeakers are not as carefully oriented as the permanent ones, they almost always create a poor acoustical response.

The contrast with Japan is informative. Both countries first experienced Western music in the sixteenth century and developed its teaching and performance in the late nineteenth century. The cultural institutions of both nations were destroyed in the mid-twentieth century: Japan's by World War II, earthquakes, and fire; China's by the Japanese occupation during that war and then by the Cultural Revolution. But Japan's fast economic recovery allowed a wave of concert hall construction as early as the mid-1970s; by Western standards many of these approximately three thousand concert halls are admirable both architecturally and acoustically. In terms of education, Tokyo's University of the Arts, including a Faculty of Music and Graduate School of Music with branches throughout the nation, has been operating since 1949. Consequently, Japan's audience for classical Western music now numbers approximately 5.6 million regular concertgoers. There are more than one thousand amateur orchestras in Japan and ten professional orchestras in Tokyo alone.

China began to build a comparable network only in the 1980s. And although it is fast catching up, it still has a way to go. There are currently eight national music conservatories in the People's Republic. These are supplemented by several thousand music education programs for all levels, from children to adults. What is now called the Shanghai Conservatory of Music was founded in 1927, but it, as well as the other conservatories, was closed for a decade during the Cultural Revolution, skipping a whole generation.

Huguang Guild Hall,
Beijing, China, 1807.

The People's Republic currently has some thirty million piano students and ten million learning violin, but there has been little ensemble playing.[11] As Alex Ross pointed out in the *New Yorker*, "China's music-education system may yield notable soloists, but it has yet to develop the breadth of talent and the collaborative mentality that engender great orchestras."[12] Still the tremendous number of young people studying Western instruments is building a more informed audience for classical music, especially in the largest cities, where there is a veritable boom in opera production, education, and composition. Today's public is for the most part unsophisticated about Western music and, apart from Beijing and Shanghai, notoriously unruly—not surprising for a culture that is used to viewing Peking opera in tea houses, where refreshments are served during performances and audience interaction with the stage is essential to the experience.

In contrast to the small, intimate tea houses and courtyards in which music has been played historically in China, Grand Theater complexes are enormous, even though the individual auditoriums they contain conform to recent trends worldwide to restrict seating capacity. Unlike the separate structures of most cultural complexes in the United States, the all-encompassing Chinese monolith feels more like a multiplex movie house than a performing arts center. Yun Jie Liu, associate principal viola at the San Francisco Symphony, speaks for many musicians when he says that "the new Grand Theaters are too big: you enter and the lobby feels like an airport."[13] But this reaction is not necessarily shared by the Chinese, who have welcomed the modern amenities of high-rise living and other megafacilities, such as supermarkets.

In fact, bigness has characterized the architecture of China's rulers for more than two thousand years, as has standardization of buildings and cities. China's current mass demolition of historic architecture, and its rapid replacement with gigantic modernist structures, has been widely criticized. However, Tao Zhu, an architect who teaches at the University of Hong Kong, points out that since the First Emperor, all great Chinese rulers have been obsessed with destroying the palaces of their predecessor and replacing them with grander ones. So while Beijing's Ten Great Buildings, built by the government in the 1950s, echo the megalomania of other twentieth-century dictators—Hitler in Berlin, Mussolini in Rome, and Stalin in Moscow—they are essentially an extension of this Chinese heritage, beginning with the Qin's Royal

Palace and Great Wall precursor (220–206 BCE) and extending to the Yuan, Ming, and Qing's Forbidden City (1406–20). Zhu sees the "building big" phenomenon that has characterized the Chinese government's public buildings since the 1990s as part of the search for an indigenous architectural identity that began in the 1920s.[14] The size of the Grand Theaters is part of this trend.

For professional musicians, a major shortcoming of the government's current policies is its emphasis on bricks and mortar, with little provision for the creation of resident companies (also a complaint in Japan). Renting to a visiting company is far more lucrative than using a theater for a resident company's productions. Chinese rental halls follow the American example of Lincoln Center (and other venues): except for the center's own festival series, troupes that perform in the various theaters when the resident company is absent pay a rental fee (as does the resident company). In large cities throughout the United States, only a handful of theaters (the Brooklyn Academy of Music and the Kennedy Center among them) pay performance groups rather than the other way around. In China, this means that the architects designing facilities for music have not had users to consult. Tan Dun, who premiered his opera *The First Emperor* at the Metropolitan Opera in December 2006 and won an Oscar in 2001 for his score to the film *Crouching Tiger, Hidden Dragon*, calls the current situation "a disaster." "China is learning fast," he says, "but it has missed the point by building concert halls that are houses for rent instead of institutions with resident companies, production budgets, and a management team."[15] Even so, visiting companies have provided a useful model for local institutions.

Indeed, once a multimillion-dollar Grand Theater is completed, in China there are no remaining funds for production or promotion. Consequently, theaters often remain dark for a year or more after they have been built. Ji-Qing Wang, a distinguished theater historian and acoustician, muses, "It is not always clear who will use the new halls."[16] The Hangzhou Grand Theater, for one, reports seat occupancy of about fifty percent, and average attendance at the Guangzhou Opera House was only seventy percent for the 280 initial performances.[17] For a culture that needs to be nurtured in Western music, the high price of tickets is daunting. The average office worker's yearly salary in Beijing and Shanghai is approximately $15,000 and in other cities approximately $9,000. In Shanghai, tickets go for an average of $50 and for star performers can rise to as much as $175.[18] But offsetting these comparatively steep prices is the considerable giveaway of tickets to politicians, businessmen, and music insiders.

In the United States, too, a donor's generosity can end with the completion of the architecture, often leaving a theater scrambling for operational funds. Richard Gaddes, general director of the Santa Fe Opera from 2000 to 2008, remarks that "new concert halls and opera houses are built as monuments to a city without sufficient thought about the money needed to maintain the buildings and sustain the programs."[19]

While China's rush to build concert halls and opera houses has some of the same virtues and liabilities of a parallel impetus in the United States, there is a big difference in numbers. China's population of 1.3 billion, with 102 cities of over a million inhabitants (compared with nine cities of comparable size in the United States),[20] magnifies every aspect of its construction phenomena. Additionally, the government's pervasive control of all projects rules out the free exchange of ideas seen in a more democratic process. Still, unfair architectural competitions, design changes (to lower costs), and the problems of financing theater maintenance and new productions are issues faced in the West as they are in China.

The descriptions in this chapter are based on a trip that began in Beijing late in 2009 and ended in Shenzhen early in 2010. With the exception of Hadid's Guangzhou Opera House, there were no architectural masterpieces. Yet as I traveled from one Grand Theater to another, what at first seemed strange became more familiar, and the questionable became more understandable.

It was initially troubling that so many Grand Theaters are separated from the fabric of the city, an apparent throwback to the now discarded 1960s idea in the United States of insulating art from the street life around it. But the rapid disappearance of pedestrian areas to make way for broad freeways in many large Chinese cities and the erection of mammoth towers near the cultural centers have given added value to the parks that afford breathing room to these facilities.

The public everywhere experiences theater primarily in the evening, and to a greater extent than in the West, those in China fare better when illuminated. Lackluster buildings such as the Grand Theaters in Shanghai and Hangzhou, and the Concert Hall in Shenzhen, come alive after the sun goes down and the lights are turned on.

A questionable aspect of these buildings is the passé science-fiction aesthetic of many of them. Paul Andreu's NCPA calls to mind a giant spaceship landing on the earth, and even Hadid's opera house has a fantastic aura. But in today's China, a country beginning to reinvent itself from scratch, futuristic symbols are understandable: having obliterated so much of its past, China is left with only the future to hold on to.

Of the Grand Theaters, Andreu's NCPA, an icon of the nation's capital completed for the 2008 Olympics, has received the most attention; the architect has the comparable Oriental Art Center in Shanghai, where Arte Charpentier's older Grand Theatre is also located. Theaters for music in regional cities include Carlos Ott's in Hangzhou; Hadid's Guangzhou Opera House; the older Xinghai Concert Hall, also in Guangzhou, designed by a group of local architects; and Arata Isozaki's Shenzhen Concert Hall. An additional project, on the cusp of completion, is a concert hall in Shanghai by Isozaki. Together these new buildings prompt conjecture about the evolution of architecture in China.

National Center for the Performing Arts (Guojia Dajuyuan)

Beijing, China
Paul Andreu, 2007
2.4 million square feet
$400 million

Plans for a national Grand Theater were first made in 1958 when Mao Zedong and Zhou Enlai initiated a building campaign associated with the Great Leap Forward. The Communist government's mass mobilization to push industrial and agricultural growth included the construction of Ten Great Buildings that were to commemorate the tenth anniversary, on October 1, 1959, of the People's Republic of China. Two buildings, one housing the Soviet-style Great Hall of the People, the other the National Museum of China and the Military Museum of the Chinese People's Revolution, erected at either side of historic Tiananmen Square, were part of this plan. The square itself was enlarged to its present colossal proportions (4 million square feet), four times its original size.

But the dramatic failure of the Great Leap Forward, with the biggest mass starvation in history (between 30 and 45 million deaths between 1958 and 1962), forced the government to put on hold several of its planned commemorative structures, including the theater. Finally, in the late 1990s the national theater project was revived by two incentives. One was a new series of Great Buildings, this time projected in

Model of Beijing.

Exterior view.

anticipation of the 2008 Olympics. The other was then-President Jiang Zemin's adoption of the NCPA as his pet project, and the possibility that he perceived as a threat the intended expansion of the Great Hall of the People toward the theater's site, which faces the Central Government's headquarters.

The importance of the site, and the feng shui/geomancy associated with it, cannot be overemphasized. The site's nearness to Tiananmen Square gives it tremendous political implications. Its proximity to the Forbidden City, the former imperial residence, confers equally important historical significance. Jiang presented the theater project as the fulfillment of Mao and Zhou's dream. Says Wang Zhengming, vice president of the NCPA and party secretary, "It was to be the symbol of China."[21]

In 1997 a closed competition to choose an architect for this prestigious commission was held; the head of Beijing's municipal government rejected the results of this process, and a second, international competition was staged the following year. Sixty-nine entries were judged by a jury of eighteen international designers and a number of city and federal government office holders. When the jury failed to reach a consensus, the Beijing government, in an unprecedented gesture, put all the submissions on public display and invited comments. Even though these opinions probably carried little influence, the exhibition established a precedent for public input that has been followed for other notable buildings, such as the National Stadium and the Grand Theater in Hangzhou.[22] Five rounds of review reduced the contestants to nine and then four finalists: Paul Andreu, Carlos Ott, Terry Farrell, and a team from Tsinghua University.

Paul Andreu is an imaginative architect/engineer who in 1968 became head of construction for the Paris airports and eventually designed the unusual tubular, reinforced-concrete terminals at Charles de Gaulle, in addition to some sixty major airports and railroad stations in Europe, the Middle East, and East Asia. The International Architecture Competition of the National Grand Theater in Beijing was one of the projects that led him to found his own architectural office in 2003. For Andreu, it was "something to clear my head," which he claims to have had little hope of winning.[23] This master of airport design was therefore elated to discover that he was a finalist for a cultural building—and at the same time dismayed by the government's decision to move the site back 230 feet from Changan Avenue (Avenue of Eternal Peace), the city's main, eight-lane east-west axis. Behind this decision was the desire to avoid conflict between the theater and either the Great Hall of the People or the Tiananmen Gate of the Forbidden City.

Andreu decided to make a virtue of necessity, suggesting an even greater setback (360 feet). He completely redesigned his submission, exchanging the classic modernist box that had brought him this far for a 151-foot-high ellipsoidal glass-and-titanium roof. Supported by welded steel trusses, the dome appears to float on a surrounding reflecting pool.

It was a stroke of genius. Of the four finalists, all of whose schemes were rectilinear, Andreu alone adapted so drastically to the change in site. The rounded form of Andreu's ellipse and its metal covering, interrupted by glazed inserts at the north and south (meant to evoke the parting of a stage curtain), are strikingly different from its neighbors and consequently defer to them.[24] The architect attributes the idea of placing the dome in water within a park to a Chinese friend's description of his first view of the Forbidden City from a nearby hilltop as "a river of gold in a green valley."[25] In a similar vein, proponents of the dome, like Wang Zhengming, offer a well-known Chinese image denoting good luck that compares its shimmering, reflective surfaces to "a pearl emerging from the water"[26] (a combination of the precious and the everyday).

Lobby; underwater passageway.

Opera house.

Anyone who has experienced the superhuman scale of the Communist rebuilding of Beijing since 1949—a gigantic reconstruction that has vastly accelerated since 1990—will understand Andreu's concern for the theater's relationship with the city. As a counterpoint to the enormous distances between buildings that discourage street life in this part of the city, he envisaged the theaters' lobbies as a more humane environment: public spaces to be enjoyed by the people. But this happened only provisionally, beginning in 2011, when $4.70 tickets for these spaces were offered.

The building's program, developed by performers, managers, and specialists of all sorts (more people than for any other building in China),[27] and overseen by government officials, called for three theaters, a recording room, rehearsal and conference rooms, an exhibition space, workshops, storage, and dressing rooms and lounge facilities for performers. Parking and technical support were to be below grade. Andreu responded with a ten-floor building, half below grade, half above. Within it, the largest and most important theater is the 2,416-seat opera, which occupies the broadest, middle area of the ellipse. Tucked into the ellipse's western curve is the 2,017-seat concert hall, mirrored at the eastern side by the 1,014-seat drama theater.

Concert hall.

Linking the three theaters within the enormous enclosing shell are vast circulation areas. Andreu had hoped that non-theatergoers might enjoy these generous interiors after descending a wide stair at the main entrance and passing through a low, glass-roofed, 250-foot-long passageway that leads under the reflecting pool (views of shimmering water are poetic in good weather, less attractive in winter, obscured by dirty ice and snow). But inexplicably, this airport expert failed to take into account the airport-like security imposed by the Chinese government at theaters throughout the country. Metal detectors at the entrance to the NCPA's underwater passage permit access to ticket holders only, either to the theaters or to the lobbies; lost was the opportunity to give the open spaces full public use. Francesca Zambello, the opera and theater professional who directed *Carmen* there in 2010, says, "The Egg [as it is popularly called] is not for the people, it's for the government. The security is deadly."[28] Still, for those working in the theater, Zambello says it is technically better equipped than anywhere else in the world.

Inside the dome a multitude of materials contrasts with the handsome sobriety of the exterior. In addition to the red-lacquer and gold-tinted aluminum and stainless-steel mesh encircling the opera house (white for the concert hall, gray for the theater), the floors are of ten different types of colored marble. Only the curved ceiling's asymmetrically placed sections of Brazilian mahogany slats bring some cohesion to the whole. A vast terrace at the uppermost level of the lobby is reserved for political receptions and the occasional exhibition. Regrettably, the deadened sound in this area, like that in the rest of the lobby, recalls the ambience of an airport rather than a festive circulation space.

And acoustics are the problem in the theaters as well. The sound quality—expected to be high in a horseshoe opera house such as this one—is compromised by padding on the rear wall and metallic mesh over the masonry side walls. The mesh is meant to be acoustically transparent but, coupled with the padding, it absorbs sound, thereby inhibiting the reflections that make for liveliness. Andreu appears to have used the mesh to mask the rectangular interior structure, retained from his original design, within which the curved theater sits. Furthermore, the control booth that extends from the rear wall of the first of the two balconies leaves a recessed area at either side that sucks the life out of any sound reaching back there. In the *New Yorker*, Ross called the sound he heard in the house's top gallery, where it should be best, "tinny and colorless."[29]

In the concert hall, Andreu's acoustical advisor, the late Jean-Paul Vian of the Centre Scientifique et Technique du Bâtiment (CSTB), a French state-owned industrial and commercial corporation, produced what Ross described as reasonably clear acoustics that lack warmth.[30] All the right elements are here: the shoebox shape that usually ensures vivid sound, two narrow balconies, a ceiling incised with irregular abstract reliefs, and a canopy over the stage. The lower level circles and flares out slightly behind the stage, giving a unity to the room. But acoustics are handicapped by the extremely steep rake of the parterre, with seats at the back almost at the same height as the first balcony, and side walls that are too far away from one another to allow the needed reflections.

Chinese opera singers and composers, in addition to the American manager Zarin Mehta, repeat the criticisms expressed by Warren Mok, the Chinese tenor and opera impresario: "The concert hall is dry, and singers can't hear themselves. The opera house is a little better but the acoustics are uneven."[31] Visiting Chinese

Plan.

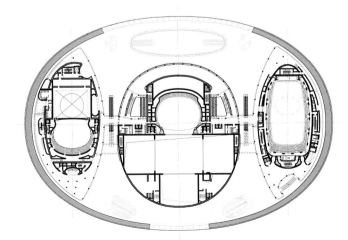

members of the San Francisco Symphony who now live in the United States experienced the same problems.[32]

The NCPA has been faulted on other counts, too. "The Egg" is the kindest of its nicknames. Local Beijingers say it is more like an upside-down wok or, inelegantly (because of its discoloration by a reddish-brown coating during spring dust storms), a floating silver turd from outer space.[33] And there is general consensus that the complex is just too big, too expensive to maintain, and too extravagant in its consumption of energy for heating and cooling.

Hao Jiang Tian, one of the few Chinese opera stars on the international stage, an imposing basso who now lives in Manhattan, recounts how lost he felt on his first visit to the NCPA: "From the front entrance it took more time to find my seat at a concert than it had ever taken me before in any of the houses I have sung in. Why so huge?"[34] The instrumentalists from San Francisco voiced a similar reservation about the vast maze of back-of-house corridors, which are spread over five basement levels and connected to the upper floors by thirty-two passenger elevators and twenty-eight escalators.

Andreu says the NCPA is "the great work of my life." Even as he continued to design and supervise major projects elsewhere, he traveled every month to Beijing for the theater's construction in what he calls a "grueling experience." One work stoppage was due to an investigation of the collapse of Andreu's cutting-edge Terminal 2E at Charles de Gaulle Airport in 2004, in which four people perished. Blame for this disaster was attributed to faulty construction methods and only marginally to the design, but the architect's reputation was irremediably tarnished.

Ironically, Andreu's expertise with airports was a liability for his design of the NCPA. The complex's enormous scale and the muffled acoustics of its lobbies are more reminiscent of a transportation hub than of a cultural center. Furthermore, Andreu's lack of experience with the architecture of spaces for music left him vulnerable to the shortcomings of his acoustical advisor. Only the architect's long-standing preference for curvilinear forms, which characterize many of his airport designs, served him in good stead for the exterior of the NCPA: the dome is a prominent landmark, its metal surface offering ravishing changes of color as the day progresses.

Shanghai Oriental Art Center
(Shanghai Dongfang Yishu Zhongxin)

Shanghai, China
Paul Andreu, 2004
430,000 square feet
$146.5 million

In 2000, competing with the NCPA (already underway in Beijing), the Shanghai municipal government selected the same architect, Andreu, to design its Oriental Art Center. The hope was that he could produce a cultural landmark similar to the capital's for Shanghai's new Pudong district. Visible across the Huangpu River from the Bund, the city's waterfront center of predominantly Western commerce, the rural area had been targeted for development by Sun Yat-sen, founder of the Chinese Republic and its first president, as early as 1919. It was to be a reclaimed Chinese metropolis that would direct attention away from what was considered to be the humiliating infringement on the city of foreign settlements.

Exterior view.

The plan for a vital Pudong district was partially implemented in the 1930s with
a few government buildings and then taken up again energetically in the late 1980s
by the Shanghai authorities.[35] Built with the same speed that characterizes all recent
construction in China, Pudong's breathtaking skyline currently boasts more high-
rises—with a dramatic array of forms—than most large American cities. It is Deng
Xiaoping's greatest achievement as the nation's leader from 1978 to 1992.

Shanghai's Oriental Art Center (SHOAC) is in every way different from Beijing's
NCPA. The capital's theater complex was conceived as a symbol of the People's Repub-
lic reminiscent of the nationalistic museum buildings of nineteenth-century Europe.
In line with China's bid for world attention with its spectacular staging of the 2008
Olympics, the NCPA signaled the country's efforts to become a major participant in
international culture.

The ambitions of SHOAC are local and national rather than global, and compared
to Beijing's inflated theater complex, a smaller size, dictated by a narrow site, makes
it more user-friendly. Both SHOAC and the NCPA stand in parks, but whereas the
Beijing dome and reflecting pool separate the complex from its surroundings, the
Pudong center connects with the exterior, even bringing within references to its syl-
van setting. The curved, greenish glass-and-steel facades connecting the cantilevered
roof with the base are supported by horizontal metal tubes that transmit stress from
the volumes of the auditoriums and isolate them acoustically. These glazed walls,
incorporating perforated-metal sheets of variable density, allow views to the exterior
and a glow at night that is more festive on the exterior than inside.

The overall form of the building is biomorphic, though Andreu denies any
thought on his part of the flower metaphor cited by Chinese commentators. In plan,
the seven-level structure resembles five petals connected to a central ovule. A first,

small petal contains the entrance hall and a steep stairway leading to a central space (the ovule) from which the theaters open in three larger petals: a 1,979-seat concert hall, a 1,054-seat opera hall, and a 330-seat performance hall. An exhibition space is located between the opera and concert halls in another small petal. There is an elaborate, under-used French restaurant on a lower level. A café squeezed into a remote part of the lobby behind the opera hall appears to have been an afterthought.

The outside walls of the different theaters, like those of the NCPA, are color-coded. Large, rounded, scarab-like ceramic tiles affixed to the exterior walls are gray for the entranceway, reddish-brown for the opera hall, and yellow for the concert hall. The public spaces are more human in scale and provide clearer orientation than those of the NCPA, but poor lighting throws a pall over the whole.

The theaters themselves share some of the problems of their counterparts in Beijing. The opera hall's wide horseshoe shape was stipulated by the client. Steeply raked central seating, upholstered in red, with slightly elevated side pods, provides good sightlines everywhere, but the acoustics—CSTB consulted here, as they had in Beijing (although Jean-Paul Vian died before the center's completion)—in the opera house and concert halls have the same lack of vitality that characterizes the NCPA spaces. Possibly because the orchestra pit, which can hold up to one hundred musicians, does not extend beneath the stage, the sound is often too strident, and several of the music professionals who criticized the acoustics in Beijing had the same reservations about the Oriental Art Center theaters.

The sound systems and loudspeakers seem to have been positioned for aesthetic purposes rather than for efficacy of sound distribution. Raucously amplifying a Chinese musical that was in rehearsal during my visit, the opera hall's sound system gives uneven coverage. Among a number of problems, the sound-control booth

Opera house.

Concert hall.

cannot be used because its glass enclosure prevents hearing what is going on in the theater. (The mixing board is, as is customary, in front of the booth.)

The vineyard plan Andreu insisted on for the concert hall, which differentiates it from the hall in Beijing, fails to realize the potential of this configuration. Lack of cohesion between the walls and the pods in which seating is located make the pods appear to have been dropped in arbitrarily. There is none of the intimacy associated with successful vineyard halls in this wide space where much of the audience is distanced from the performers. The sound was uneven and had little punch at a New Year's Eve concert of popular Viennese-style music, a problem attributed by Ji-Qing Wang, the Chinese acoustician, to thin wood wall panels that absorb sound because there is space behind them.[36]

Two leading theater-management groups, the Wenhui-Xinmin United Press Group and the Poly Culture and Arts Co., Ltd. (a division of the Poly Group, founded in 1993 as part of a huge, party-controlled consortium of military, real estate, and engineering concerns, among other industries), took over the Shanghai Oriental Art Center in early 2004. They engaged a group of music experts who made more than two hundred design improvements to the theaters. Li Nan, head of the Poly Group, blames the problems they found on the lack of consultation between the architect and the users.[37]

Shanghai Grand Theatre (Shanghai Dajuyuan)

Shanghai, China
Arte Charpentier, 1998
Over 590,000 square feet
$150 million

In 1994, when government officials in Beijing were still only talking about erecting a new theater, they were upstaged by their counterparts in Shanghai with the announcement of a competition for a new opera house. Like Andreu, who was working on the Pudong Airport (1996–99) when he entered the NCPA competition, the late Jean-Marie Charpentier, who won the commission for the Shanghai Grand Theatre, was already involved in a major project in China: the plan for Pudong's

Exterior view.

Lobby; Lyric Theater (opera house).

immense (430,556 square feet) City Hall Plaza (1996). This urban plan was followed by, among Arte Charpentier's many other projects in the People's Republic, a design for the eight-lane, three-mile Century Avenue leading east from the Oriental Pearl TV Tower near the river to the plaza and, just before the plaza, to the park where the Oriental Art Center is located.

In contrast with Andreu's outlying project for Pudong, Charpentier's Shanghai facility is smack in the historic center of the city. It neighbors People's Park (formerly occupied by the British Jockey Club's race course, which was converted by the Communist government into a giant ice-skating rink), in which the Shanghai Museum (designed by the local architect Xing Tonghe) opened in 1996.

With its population of twenty-two million, Shanghai is China's largest city and the nation's only one with a Western-acculturated audience used to buying tickets for the performing arts. Indeed, various foreign opera companies visited the city regularly between 1883 and the 1920s, and the Shanghai Public Band became a symphonic orchestra in 1907. Despite political and wartime interruptions, the orchestra developed into the Municipal Council Symphony Orchestra in 1922, promoting Western music and Chinese orchestral work. In 1956 the large group of musicians became today's Shanghai Symphony Orchestra for which yet another concert hall (designed by Arata Isozaki) is currently under construction.

The architecture of the ten-story (of which two are underground) Grand Theatre tries to combine historic Chinese architecture and modernism in the same way that Le Corbusier and Louis Kahn incorporated indigenous themes within modern architecture at Chandigarh and Bangladesh. But the attempt is less successful here (or in other such couplings, like Carlos Ott's Hangzhou Grand Theater). The Grand Theatre's roof appears as a holdover from the 1950s when Le Corbusier and Eero Saarinen used similar forms to better effect.

Despite its shortcomings, the Grand Theatre's half-moon roof is generally liked by the Chinese, for whom it symbolizes the sky, and copies of it are found all over China.[38] The building's square plan also resonates in this nation, where the shape traditionally refers to the earth. And the steep stairway up to the entrance doors is preferred to the downward entrance stairway (associated in China with tombs) of Andreu's complex in Beijing.

In its favor also is the transparency of the building's glass curtain wall, suspended from cables, which anticipated the potential for openness in Chinese society, meta-phorically speaking at least. Immediately behind the facade, prominent freestanding structural columns recall the Chinese fondness for this architectural element. The lobby, with walls of white marble trimmed with black and yellow stone and a

Plan.

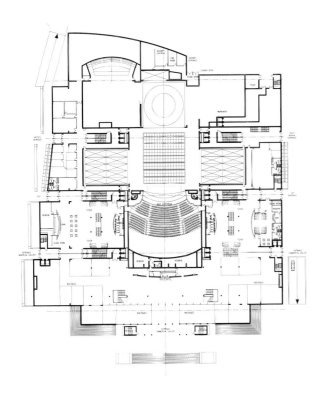

white-marble double grand stairway (designed by the United States firm STUDIOS), is elegant and festive. This area is dazzling in sunlight; at night its bright lights cast a welcoming luminosity to the exterior.

The horseshoe-shaped, 1,800-seat opera house (called the Lyric Theater) is used for Western opera, ballet, musicals, and symphonic concerts. There is also a six-hundred-seat theater for Chinese opera and chamber music, a two-hundred-seat experimental space for plays, and several rehearsal rooms, in addition to backstage support spaces. A glamorous rooftop restaurant is popular, as is the Maxim's de Paris licensee at street level.

The walls of the Lyric and the smaller theaters are light-colored wood; these become open grilles at the upper and rear parts of the opera house to hide movable sound-absorptive curtains. The Lyric's two balcony circles suggest a traditional opera house, but three tiers of boxes at either side of the stage have such poor sight lines that their seats are not normally offered for sale.

Generally, the acoustics of this hall (L-Acoustics served as acoustical consultants) are considered slightly better than those of the Shanghai Oriental Art Center. However, several musicians have described the theater's sound as undistinguished.[39] At conductor Riccardo Muti's 2010 New Year's Eve concert of overblown, brassy, light classical music, the acoustics were indeed disappointing. Big clashing chords were loud, but without any sense of envelopment or excitement, and the orchestra did not sound blended. Distracting video screens at either side of the stage alternated photographic images with program explanations. Yet despite being something of a mixed collection of design approaches, compared with many of the theater complexes that have followed it in China, Charpentier's building has withstood the test of time.

Hangzhou Grand Theater (Hangzhou Dajuyuan)

Hangzhou, China
Carlos Ott and Petroff Partnership Architects, 2004
540,000 square feet
$112.5 million

Carlos Ott's streak of winning commissions for Grand Theaters is surprising. The Uruguayan-born architect, roundly vilified not only for the mediocrity of his Bastille Opera House design (1989) in Paris but also for his arrogant attitude throughout the project, has built more theater complexes in China than any other single practitioner: four up and running and a fifth under construction.[40] Ott's success in this country reveals a lot about differences in taste and priorities for theaters between China and the West, together with the remarkable extent to which the French continue their historic influence in this country.

Although Ott is not a French national (he is now a Canadian citizen), he is identified with French architects because of his role at the Bastille Opera House, which

Theater with International Conference and Exhibition Center at left, 2004.

Exterior view.

is greatly admired by Chinese officials who have visited it. The year the Paris house was inaugurated, Ott was invited by the Chinese government to participate in the competition for the Grand Theatre in Shanghai (an invitation he turned down, since he was busy working as a partner with the NORR Partnership, Ltd., Canada's largest architectural, engineering, and planning firm).[41] Subsequently, his premiation as one of four finalists in the Beijing NCPA competition apparently encouraged him to enter other opera house competitions in China, all of which he won: Nanjing (1996), Hangzhou (1999), Wenzhou (2001), Dongguan (2001), and Zhengzhou (2003). The Hangzhou Grand Theater (completed in 2004) exemplifies many of the characteristics of Ott's theaters that have endeared him to the Chinese: workability, affordability, and the building as metaphor.

Hangzhou, with a population of over six million, is located about 120 miles southwest of Shanghai in an area of rich farmland; it is one of the oldest and most picturesque cities in China. The locale has always enjoyed a near mythical status for its beauty and as a place of retreat in troubled times. It is situated beside the famous

West Lake, whose tranquil, willow-lined banks are dotted with ancient pagodas, beyond which misty hills rise in a landscape reminiscent of classic Chinese brush paintings. So enchanting is this historic city that it inspired the saying "Above there is heaven, below there is Hangzhou and Suzhou."

Originally, the Grand Theater was to have been built near West Lake. For his building, the architect repeated the metaphor of a pearl emerging from the waters, intended as a symbol of the city. But a 1999 feasibility study for the complex revealed that the designated downtown site was too small, and its design was thought to conflict with nearby buildings. In 2001 the project was moved to its present location east of Hangzhou near the Qiantang River in one of the nation's many formidable high-rise New Towns. The Grand Theater is best seen from across the river, beside which its graceful curve stands out. Even this clear view, however, is compromised by the vulgar, gold-colored, spherical International Conference and Exhibition Center, also by Ott, that faces the Grand Theater on the same shore. Neither structure is helped by the backdrop of unsightly towers in the new civic center, completed in 2011.

The program for the competition—with a jury consisting exclusively of citizens of the People's Republic of China—was compiled by technical specialists, government officials experienced in construction, and a planning commission whose main goal was to produce a popular landmark. In order to achieve that aim, Hangzhou followed the lead of the NCPA, exhibiting to the public models and explanatory texts provided by the three competition finalists: over twenty thousand votes were cast, with a majority for Ott.

Opera house; banner reads "2010 Hangzhou New Year's Concert."

Concert hall.

Despite the switch away from West Lake's gentler environment, the architect retained the poetic concept he had developed for that area. The 1,600-seat opera house, 600-seat concert hall, and 400-seat multiuse hall are contained within a titanium-covered, crescent-shaped, prestressed-concrete structure that embraces a glazed truncated cone: a pearl within a shell. For acoustical reasons, the two planes of the exterior wall are separated by four inches, as are comparable double floors. Ott's later scheme for the sphere of the conference center symbolizes the sun. A reflecting pool between the two buildings represents West Lake; the composition is oriented toward the pool and has views to scenic Baoshi Hill.

Circulation, confined to an ample, predominantly white, day-lit space running the length of the concave glass wall of the "shell," provides clear access to the three venues. The opera house is a fairly wide horseshoe. The two deep balconies at the rear do not wrap around in a traditional arrangement, except for two shallow side balconies designed to look like boxes stepping down. Seats in these areas face across the room instead of toward the stage, and their sight lines, like those from the back rows of the top balcony, are severely hampered.

The house's walls are for the most part painted red; black floors and balcony facings provide the remainder of the theater's color scheme. (The management added gold swags to the balconies to temper what it considered to be excessive black.) This auditorium will not win any prizes for aesthetics, but based on a rehearsal without an audience of Western classical and modern Chinese music performed by the eighty-piece Hangzhou Philharmonic Orchestra, the acoustics, credited to Müller-BBM, are good. Certainly helped by the fact that the hall was empty, sound came across as full, rich, and warm. Compared with many of the newly built spaces for music in China, the sound was also more enveloping.

The birch-composite-board-paneled concert hall is more appealing than the austere opera house. Pale wood seat backs and frames (with purple upholstery) harmonize with the walls and floors. With seating (and a large pipe organ) that extends behind the stage, the rectangular room is intimate and refined. Only the continental seating of the fairly long rows (twenty-nine seats per row)—necessitated by the tight site—feels a bit cramped.

Two rehearsals—again in an empty hall—were particularly interesting because one was played on a Chinese instrument, the other on a Western one. The delicate sound of the seven-string, zither-like qin, an instrument of ancient origin, was easily heard, with clarity and sonority, as were the works of a brief piano recital.

Along with the impressive technology of both theaters, power-driven lifts and movable chairs in the small multifunctional hall allow an exceptional number of reorganizations. There are also two rehearsal rooms. A Western-style restaurant is planned for the top floor. Built into the reflecting pool in front of the complex is a seven-hundred-seat amphitheater. It is part of a large landscaped area around the Grand Theater and the conference center that mercifully sets them apart from the surrounding megastructures. In this sense, Ott's Hangzhou Grand Theater is different from cultural complexes that are overwhelmed by gigantic construction nearby—as, for example, the opera house in Guangzhou by Zaha Hadid.

Plan.

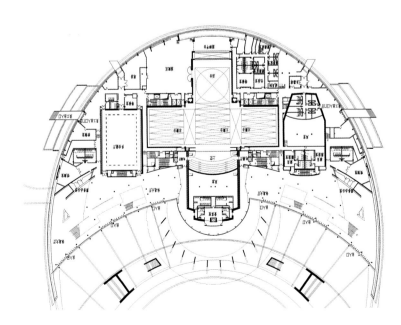

Guangzhou Opera House (Guangzhou Dajuyuan)

Guangzhou, China
Zaha Hadid Architects, 2010
495,000 square feet
$202 million

Leading China's economic revolution and the extraordinary construction campaign it has spawned is the Pearl River Delta in the country's southernmost Guangdong Province. There were good reasons that Deng Xiaoping designated this subtropical area for experimentation in 1978: the Cantonese-speaking inhabitants of the immense region have a history of more than two thousand years of fierce independence, skillfully trading with Southeast Asia, the Middle East, and by the mid-sixteenth century, Europe.

Guangzhou, in Guangdong Province about one hundred miles northwest of Hong Kong, is a vibrant trading city, once known to the West as Canton, that traces its history to the third century BCE. In 1984 it was named a Special Economic Zone (SEZ), one of fourteen open coastal cities so designated at the time. These earlier treaty ports conceded to foreign merchants were targeted as agents of Chinese renewal, where foreign capital would prime China's economy.[42] With a current population of nearly ten million, Guangzhou has become an important center of foreign commerce in South China, producing approximately one-eighth of the world's industrial output.[43] Its link to Shenzhen, the initial SEZ city, by the first high-speed, limited-access highway in China (opened in 1994), is the lifeline of this economically vital region.[44]

The Guangzhou Opera House is not located within the original city limits, however. Rather, like the Grand Theater in Hangzhou, it anchors a new central business district, this one known as Zhujiang (Pearl River) New Town, on the edge of the urban center. Consisting as recently as 2006 of rice paddies, the area has become a megacity of colossal proportions: the Guangzhou Twin Towers rise more than 1,400 feet, the height of the Sears Tower in Chicago, and its TV Tower, from which the Asian Games were broadcast in November 2010, is more than 2,000 feet.

As for all such developments in China, a master plan was drawn up, here by the Obermeyer Group, German engineers and architects who gave what was billed as "the leading opera house in the world" a place of honor at the northwest end of a cultural park. At the park's opposite end is the forbidding masonry Guangdong Provincial Museum (designed by Rocco Yim, 2010); the Guangzhou New Library (Nikken Sekkei LTD, 2010) and the Guangzhou No. 2 Children's Palace (SBA International, 2005), both

unremarkable, are at the north side. A broad north-south promenade separating the
Opera House and the other cultural buildings is the new city's axis.

London-based Zaha Hadid designed two powerful monoliths for the complex
that struggle to hold their own in relation to the high-rise construction surrounding
the park. They are helped by the grass-covered, artificial hillocks with which Hadid
has filled the area overlooked by the Opera House, and which create a scale sympa-
thetic to her architecture. The two-level complex is entered by car at the north one
level below grade between the two buildings (there is a 250-space parking facility)
and on foot from several directions, notably at the east from the park. A sinuous
ramp swoops up to an elevated pedestrian platform that brings visitors to the
entranceways of the 1,804-seat Grand Theater and the 440-seat multiuse hall. A café,
restaurant, and retail facilities are embedded in the lower level, accessed from the
park and by escalators from the lobbies of both theaters.

As they ascend the ramp, theatergoers are immersed in a world of amorphous,
gently layered forms, set apart from the New Town. The computer-modeled dual

Zhujiang (Pearl River)
New Town with
Guangzhou Opera
House across river at
left, rendering, 2010.

structures are sited close to one another: a larger black-granite volume contains the Grand Theater, and a smaller white-granite building, standing beside a large reflecting pool, contains the multiuse hall.

The shells of the two constructions consist of concrete-and-steel superstructures that are supported and braced together at some high points. In the large volume, glass held within the steel frame opens the walls and portions of the ceiling to views of the park. Unique starlike cast-steel junction boxes maintain the structure's tension. Steel filigree recalls the metalwork of Herzog & de Meuron's widely publicized Beijing National Stadium (2008), or "Bird's Nest," but Hadid's is much more transparent.

The opera house's fly tower is completely hidden within the building, giving the exterior a massive quality, shared by the smaller building. The black-granite surface of the exterior flows inside to become the floor of the majestic, column-free lobbies, from which sweeping staircases rise overseen by balconies that appear to have been sculpted from the theater's enclosing wall.

Within the theater's dynamic, seamless space, two cantilevered, curved balconies are designed as undulant terraces. The contours of the golden walls have a smooth Corian-like appearance, as do the white balustrades of the stalls and the balconies. In reality, the walls are clad in stiff glass-fiber-reinforced gypsum, out of which half-inch-thick scoops were gouged to diffuse sound. Four thousand tiny LED lights poetically stud the concave ceiling and underside of the balconies. Hadid feels that the theater's "balanced asymmetry" enlivens the space visually and may even give it an acoustical edge over strictly symmetrical theaters.[45] (The acoustical consultant was Marshall Day Acoustics in Australia.) Her conviction is borne out by *Financial Times* music critic Ken Smith, who cited the "superb acoustics" in his review of the house's opening night performance of *Turandot*. A year later he noted that adjustments to the hall "offered a finely layered acoustic where the orchestra, soloists and choristers

Opera house looking
toward stage and
toward balconies.

from the China Opera and Dance Theater could each blossom without interfering with one another sonically."[46] The composer Huang Ruo calls the opera house "the best for orchestral music of any auditorium in China, Taiwan, and Hong Kong."[47]

Two dance rehearsal rooms develop the wall treatments of both theater and multiuse hall. A solid-surface material (HIMACS by LG), resembling the opera house walls, is used on the lower areas (here the incisions are made absorptive by fabric behind them), while plaster on the upper areas is molded into graceful reliefs that recall rippling water, an important acoustical feature that breaks up reflections. Aesthetically, the artificial light that shines through the spaces between the reliefs contributes to the stunning appearance of these rooms.

In contrast with the unconventional opera house, the multiuse hall is a shoebox, but a regular grid of adjustable floor modules provides great flexibility for various configurations. Rows of seating mounted on each unit can be raised, lowered, and rotated mechanically; the rows, which can be stored under each module when not in use, are positioned manually. The hall is surrounded by a billowing white-plaster foyer.

The architect conceived the two structures as twin boulders from the nearby river, but in reality the complex is more evocative of spaceships or stealth bombers. The Opera House is billed as "overlooking the Pearl River," and those views were considered a major feature of the site. Unfortunately, the four-lane Linjiang Avenue

Multipurpose hall.

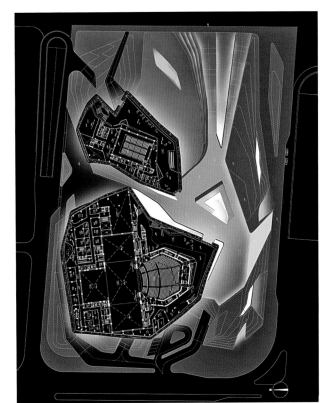

separates the park from the water, and the huge TV Tower, and the many other
unsightly skyscrapers that have risen on both sides of the river, detract from the
effect of the handsome cultural buildings. An additional encroachment on the
opera house is the subway line and station, which are much too near the theaters for
comfort, even though China's modern subways are relatively quiet compared with
American ones.

Hadid's Opera House stands alone in China for the innovative forms of both the
complex as a whole and the Opera House interior. It is regrettable, however, that like
so much Chinese construction, poor detailing compromises the building's overall
quality (early in 2011 some of the most carelessly affixed panels were replaced). As in
the case of her winning entry for the Cardiff Bay Opera House in Wales (1995), albeit
designed quite differently, Hadid has created an exceptionally generous and inviting
public domain. At Guangzhou the asymmetry of the Opera House interior realizes
a concept that went unbuilt at Cardiff, while the structure as a whole reflects the
impressive biomorphic forms of her most recent work.[48]

Shenzhen Concert Hall (Shenzhen Yinyueting)

Shenzhen, China
Arata Isozaki & Associates, 2008
603,651 square feet
Concert Hall: $200 million

Shenzhen, a few minutes' drive from Hong Kong, was the first and most successful of four cities in Guangdong Province that were designated SEZs in the early 1980s. At that time a fishing village of less than thirty thousand, Shenzhen is now a metropolis of about fourteen million. Among the city's many current construction projects is Rem Koolhaas's Shenzhen Stock Exchange of more than two million square feet—almost twice the volume of the Empire State Building—which is meant to challenge Shanghai's stock exchange.

In 1998 Arata Isozaki won the international competition for the design and construction of the city's concert hall and library, two parts of the Shenzhen Cultural Center. Isozaki placed his three-story poured-concrete-and-steel buildings on a twenty-foot-high platform in order to set them apart from the spatially incoherent buildings towering around the site. The platform, from which the main entranceways are accessed, unites the library (at the south) and concert hall (at the north) with a

Library (at left) and concert hall.

Lobby.

bridge over the east-west street that runs between the two facilities. Views underneath Lobby. the platform to the street beyond give some transparency to this dense area.

Within their unruly setting, Isozaki's elongated, low-rise twin structures appear restrained and dignified. However, at the time they were built, in an early stage of China's race to modernize, construction skills were still inadequate to successfully detail the undulating, gridded, glass curtain wall at the east of both the library and the concert hall. The west facades' solid granite walls were perhaps meant to emphasize the symmetry of the urban design, but they render this side of the complex lifeless.[49] All in all, Isozaki's buildings do not achieve the kind of daring that characterizes this architect's best work.

In their favor are the identical transparent entranceways facing each other across the elevated terrace. Their roofs, series of spiderweb-like polygons composed of glass-filled steel triangles in different sizes and shapes, step down gracefully from the top of the structures to the glazed entranceways, sending a powerful signal both in daylight and when illuminated from within.

Entered at street level or from the platform above, the library and the concert hall's foyers continue their exterior dialogue with similar giant, angled, structural pillars finished in silver leaf for the library and gold leaf for the concert hall. These grandiose spaces, traversed by escalators, stairways, and bridges that lead to the programmatic spaces within, are dramatic urban rooms, eminently appropriate to the cultural activities they introduce. Additionally, the informal stage in the lobby of the concert hall has attracted young people and made events and performances more accessible to the public.[50]

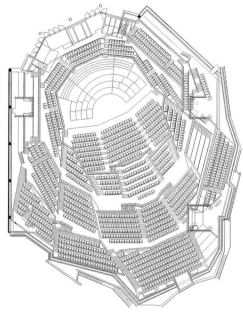

Concert hall; plan.

The concert hall contains a 1,680-seat Symphony Hall and a modular, 400-seat Theater Studio, plus amenities that include rehearsal rooms, a café, and a regional Chinese restaurant. The hall's surround seating is the product of Isozaki's close collaboration with Yasuhisa Toyota. They had worked together earlier on the Kyoto Concert Hall (1995), a handsome building whose shoebox auditorium was described by the critic Paul Goldberger in the *New York Times* as first rate, "blending a sense of intimacy and monumentality."[51] The hall's acoustics received comparable praise from the acoustician Leo Beranek.[52]

Having triumphed as the acoustician for the vineyard-shaped Suntory Hall in Tokyo (Yasui Architects & Engineers, 1986), Toyota persuaded Isozaki to adopt what he believed to be visually more intimate surround seating for Shenzhen.[53] The city representatives who served as clients were easily persuaded to accept the asymmetrical, terraced seating configuration that rises around the stage. The stage itself has thirty-six elevator platforms that can be adjusted to the needs of the performers.

Slightly angled sound-reflecting boards are placed along the auditorium's upper walls and ceiling. The bright red seating arrangements neither constitute a true vineyard, like that of the Berlin Philharmonie, nor innovate, as do the seamless overlapping terraces of Hadid's Guangzhou Opera House or Jean Nouvel's project for the Paris Philharmonie (see Chapter 5). But they are effective acoustically. Additionally, a heavy concrete ceiling provides the mass needed for acoustic reflections. Repeatedly, musicians who have experienced the Shenzhen Concert Hall praise the fine quality of its sound. Having conducted there twice, Tan Dun declares the hall "amazing architecturally, with fantastic acoustics."[54]

Xinghai Concert Hall (Xinghai Yinyueting)

Guangzhou, China
Architects/Teachers at South China University of Technology,
Architectural Design Institute, 1998
19,375 square feet
Cost undisclosed

Built in 1998 as the home of the Guangzhou Symphony Orchestra, the modestly scaled Xinghai Concert Hall is a unique example of Chinese architecture conceived for a resident company. Named after Xian Xinghai (1905–45), the famous patriotic composer and Communist party member, it has a 1,437-seat auditorium—China's first vineyard-shaped concert hall—and was modeled after the Berlin Philharmonie. Among musicians familiar with China, the hall is well known for its fine acoustics. The building also contains a 462-seat chamber music hall, a 96-seat multimedia facility, and several rehearsal rooms. The completion in 2009 of a support building across the street demonstrated governmental sponsorship of the Guangzhou Symphony Orchestra. This addition contains individual practice rooms for orchestra members, more rehearsal space, and offices.

Located on captivating Ersha Island in the Pearl River, the Xinghai Concert Hall enjoys a much better relationship with that body of water than does Hadid's Opera House. At first sight, the hall's roof (a hyperbolic paraboloid known as a saddle surface) brings to mind the similarly shaped roof of Le Corbusier and Iannis Xenakis's Philips Pavilion (1958) and the Gare de Saint-Exupéry TGV airport railway station in Lyon (1994) by Santiago Calatrava. Gridded glass-and-steel facades drop at either side of

Exterior view of concert hall (at left) and support building.

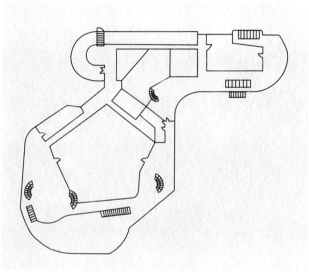

Concert hall; plan.

the building from the highest points of the reinforced-concrete, saddle-shaped shell, enclosing the lobbies and allowing views to the subtropical landscape beside the river.

Materials and the shape of the building were chosen for acoustical excellence. The main hall is wide, but the concave reinforced-concrete ceiling, walls of hardwood attached to plywood, fixed sound-reflecting canopy over the stage, and solid wood floors all contribute to a good tonal balance and evenly dispersed sound. A New Year's concert there featured the soprano Wu Bixia. Her fine voice was amplified loudly but with good localization; the sound clearly came from the singer, not from where loudspeakers were located.

The multiuse chamber music hall also serves theater, conferences, and film. To allow variable acoustics, twenty-nine revolvable cylindrical tubes, one side sheathed in beech plywood and the other in a sound-absorbing material, line the side and rear walls of the irregularly shaped room. Beech plywood is used for the other walls. Additionally, there are adjustable, sound-absorptive curtains. Although the designers addressed acoustic flexibility, the hall's low ceiling may contribute to the somewhat dry acoustics, limiting its effectiveness for chamber music.

Architecturally, the exterior of the Xinghai Concert Hall is too derivative to be noteworthy. But its main hall offers excellent acoustics and the visual intimacy lacking in many of China's Grand Theaters, such as the Shanghai Oriental Art Center's concert hall. Perhaps the growing number of talented Chinese architects will produce other grass-roots-initiative projects like the Xinghai.

WHATEVER THEIR FAILINGS, CHINA'S MOST RECENT MUSIC VENUES ARE AN improvement over most of their predecessors. Late-twentieth-century concert halls in Beijing have problematic sight lines, lighting, and acoustics and offer few amenities. The Poly Theater (1991), for example, shares an undistinguished lobby with a hotel and shops, and its big uncovered orchestra pit makes it difficult to achieve balanced music. The Forbidden City Concert Hall, a roofed-over open-air venue, has in its favor mainly old-fashioned charm. The Shanghai Concert Hall, a 1930s movie theater renovated in 1959 (and moved more than 164 feet to a nearby location in the course of a 2002 facelift), fares somewhat better.

The seven music theaters described here are but a small part of this building type in China in the last decade, and an even smaller part of what is projected for the future. Two major complexes, both completed in 2007, are Paul Andreu's Science and Cultural Arts Center, including a Grand Theater and dinner theater, in Suzhou and the Chinese architect Huang Jie's Qintai Grand Theater in Wuhan. Among the many recently completed Grand Theaters, Chongqing's (2009) is one of the most ambitious, apart from Beijing's, in terms of size. The complex (one-million-plus square feet), by the German architects Meinhard von Gerkan and Klaus Lenz, stands beside the Yangtze River in China's most populous municipality (over thirty million people, including suburbs and a large rural hinterland). And in 2011, the Shanghai Culture Square (Beyer Blinder Belle with Zhang Congbo) was inaugurated. Built underneath People's Park, the performing arts complex includes a two-thousand-seat theater intended for commercial musicals but allows for some classical music in the programming.

In his fascinating history, *Architecture of Modern China: A Historical Critique*, Jianfei Zhu points out that Western-style architecture arrived in China in the seventeenth century at approximately the same time that Western music was introduced by Matteo Ricci. Two centuries later Western architecture reappeared in the modern urban facilities of the treaty ports, peaking in the 1930s and 1940s. But Communist ideological prejudice against the International Style, regarded as an imperialistic tool, slowed the development of modernism in China, restricting the style to its economic, functional, and technological aspects rather than allowing innovation, especially in the capital.[55] Not until the late 1990s did more Chinese architects—many educated in America, Europe, and Japan—break with the decorative Social Realist style to adopt purist and tectonic modernism.[56] The intermittent pattern in exposure to forward-looking modern architecture in China is paralleled by the lack of a continuous tradition of Western music.

As the built environment in the West illustrates, familiarity with well-designed buildings is no guarantee of consistent architectural excellence. But such knowledge can provide guidelines for both the public and the client. Currently, just a few years have passed since the first dramatically inventive Western buildings appeared on the scene in the People's Republic. And it is only since 1990 that young Chinese students and practitioners have begun to travel abroad and to have access to professional literature, both essential to the development of their own ideas. Yet Jianfei Zhu already sees a growing group of talented Chinese architects who are transcending mainstream conventions, and he, in agreement with Rem Koolhaas of OMA, considers modernization to be most intense in Asia.[57] Bolstering these observations is the fact that important Chinese-born architects have positions at two major institutions in the United States: Qingyun Ma is dean of the architecture school at the University of Southern California, and Yung Ho Chang was, until May 2010, chair of the architecture department at MIT.

Having built in China since the mid-1990s, Arata Isozaki sees considerable improvements in construction expertise. For his first project in China, the Shenzhen Concert Hall, he was surprised that the Beijing architects associated with the project, although proficient with concrete, did not know how to make working drawings for the steel structure. Construction also suffered from the laborers' inexperience: despite regular site visits, Isozaki finally gave up on the quality of the curtain wall, which he describes as "terrible by Japanese standards." But by the time he completed the Central Academy of Fine Arts Museum in Beijing (2004–8), he saw that the young local architects and engineers had learned to control the site.[58] Yet as evidenced by the problems at the recently completed Guangzhou Opera House, the quality of construction in China still has a way to go.

Yasuhisa Toyota, Isozaki's acoustical advisor, had a similar learning arc in Shenzhen regarding communication with the client. He describes as extremely uncomfortable his initial meetings with older bureaucrats for whom the presence of experts was not helpful. These first contacts are reminiscent of the lack of meaningful exchange between architects, acousticians, and other technicians and the clients and users at Lincoln Center in the mid-twentieth century; communication improved in Shenzhen only when successors to the politicos came into the picture, users and managers who had a better understanding of the project, Toyota says, and "learned a great deal from it."[59]

Paralleling increased design and construction expertise are similar trends in China's music world. Among the many permanent symphony orchestras formed in China in the early 1950s, several of those still active have full seasons, like orchestras in Europe and the United States. The Shanghai Symphony, the China Philharmonic, and the Guangzhou Symphony—all under the direction of Long Yu, who is fast becoming as much of an international star as was the Berlin Philharmonic's Herbert von Karajan—are examples. Hangzhou boasts recently created provincial and city orchestras. Early in 2010 Beijing's NCPA joined the movement, forming an in-house

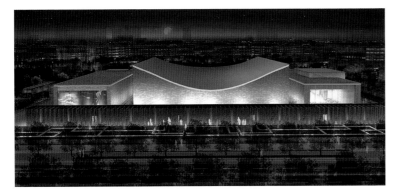

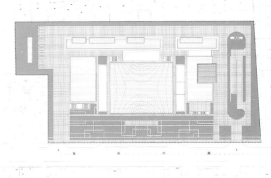

symphony orchestra. Since about 2003 a number of music festivals have sprung up, including one in Guangzhou and two in Beijing. Paradoxically, Chinese orchestras are being founded at the same time that long-established symphony orchestras in cities across the United States—among them Cleveland, Philadelphia, Pittsburgh, and Detroit—are undergoing traumatic changes for economic reasons, and others— the Florida Philharmonic and the Honolulu Symphony, to name but two—have gone out of business.

If the West's alternatives to mainstream concert halls and opera houses are a measure of public sophistication, the Chinese are headed in the right direction. In recent years, a number of private alternatives to government-sponsored venues have come into being. A small auditorium devoted to film screenings, lectures, and contemporary music in Beijing's Ullens Center for Contemporary Art is typical of the industrial buildings that are being converted into spaces for music and theater.

Under the guidance of Tan Dun, several Ming-era houses in outlying Zhujiajiao, a famous old water town in Shanghai's western suburbs, were renovated by Isozaki to create spaces for ancient Chinese music; operations began in 2010. Tan's project to preserve disappearing traditions includes performances both indoors and out, presented with newly designed lighting effects—concepts he hopes can eventually be expanded nationwide.

Theater management is also becoming somewhat more diversified with the addition of the China Arts and Entertainment Group, closely aligned with the Ministry of Culture, participating in the management of the NCPA and the Grand Theaters in Shanghai and Guangzhou. After an unsuccessful trial with a German team, the Grand Theater in Hangzhou has taken over its own management. Until recently, theaters were run only by appointees of the municipal government or by the Poly Group. The Poly Group continues to dominate performing arts management in China, relying on its connection with the People's Liberation Army for the clout needed to be effective in China, and using its theater expertise for the construction and development of its own theaters.

A promising project under construction in the historic former French concession area of Shanghai is a 215,000-square-foot concert hall designed by Isozaki with Toyota for the Shanghai Symphony (slated for completion in 2013). Computer renderings show restrained architecture characterized by a gently concave roof, a subtle half-moon in form. In order to fit the 1,200-seat main auditorium and 400-seat chamber music and recording hall into the tight site, four floors will be below grade, two above. The concert hall itself will resemble this team's successful Kyoto Concert Hall, in which limited seating is placed around the stage.

The project faces, among other hurdles, noise and vibration from a nearby subway, necessitating a special ceiling and side walls. But Toyota points out how different this experience will be: "Working with Long Yu as the client is a superior situation to Shenzhen, where there was no orchestra with whom to communicate."[60]

Hong Kong's West Kowloon Cultural District is the biggest such enterprise in this part of the world. In 1998, the local government initiated the project for a number of venues to be constructed on approximately one hundred acres in reclaimed land across the bay from the city proper. A master plan for the district, awarded early in 2011 to Foster + Partners, enjoys a $2.8 billion endowment. Norman Foster is a familiar name in Hong Kong, where he was the architect for the HSBC headquarters (1986) and the Hong Kong International Airport (1998) and the Cathay Pacific lounges there (2012). His proposal for a large park, cultural institutions, art education facilities, and commercial facilities enjoyed a slight edge over the two other competitors (Rocco Yim and OMA) in a public poll held before the results were announced.

The most intriguing question in regard to new architecture in China is the future of indigenous practice. Many Chinese music facilities are based on Western models: the tripartite Grand Theaters on North American cultural centers, the theaters themselves on Western precedents. With few exceptions, such as the Xinghai Concert Hall (designed by Chinese architects in imitation of a 1950s French pavilion), the architects and acousticians for these venues are also Western (or in the instance of the acoustician for Hadid's building, Australian). Does the opportunity now present itself for Chinese architecture to break with the two-century-long domination by Western ideas? It was only in about 2001 that the government ended control of all architecture by the Design Institutes, allowing the development of individual practices.

An important related issue is what will be performed in these spaces. Among the many events served by China's Grand Theaters are political gatherings and political performances, both of which are about pomp and circumstance, not art and acoustics. Western music—whether Soviet-style extravaganzas, Broadway shows, or Italian-style grand opera—dominates current programming, with a few slots given to historic

Chinese instruments, which in solo performances are always amplified, usually badly. Only full Chinese orchestras (about seventy musicians playing modernized Chinese instruments, as the Hong Kong Chinese Orchestra does) escape general amplification.

Chinese composers, however, are influenced by the traditional instruments of their own country and are writing music that combines Eastern and Western elements in novel ways. An older generation includes Zhou Long, Qu Xiao-song, Guo Wenjing, Chen Yi, Tan Dun, Bright Sheng, and Bun-Ching Lam (who was born in Macau); among the younger generation are Du Yun and Huang Ruo. (The enormous increase in the number of music students in China must also be a factor in the development of classical music.) In order to be successful, the limited amount of non-amplified Western music currently programmed in new theaters requires that these theaters have higher-quality natural acoustics in the spaces that are shared with epic extravaganzas.

The 2008 financial meltdown in the United States and Europe manifested the West's own kind of destabilization. Even as China enjoys a period of unprecedented prosperity, the country faces future upheavals as social dislocation, social closure, and the marginalization of certain groups become increasingly apparent.[61] (The angry protests of citizens who lost children in the 2008 Sichuan earthquake because of faulty school construction are just one such mass anti-government demonstration.) How this volatile social and political situation will affect the country's cultural life is impossible to predict.

5 The Future
Near and Far

A

LL ARCHITECTURAL ENDEAVORS HAVE THEIR RISKS and setbacks, but opera houses and concert halls, with their uncompromising acoustical requirements—including the enormous physical volumes needed to fulfill these requirements—and complex issues of access and egress, seem to be particularly prone to reversals. Long before Frank Gehry's Guggenheim Bilbao was completed in 1997, Jørn Utzon's Sydney Opera House (1973) had established an admired precedent for popular destination architecture. But what became a familiar international landmark took sixteen years to build and cost $102 million, more than ten times the original estimate. Nor did the building's problems end with its realization: despite years of costly acoustical adjustments, neither the opera house nor the concert hall has ever functioned satisfactorily.[1]

A century earlier, it took twelve years to erect Paris's Palais Garnier, an undertaking that many wanted to abandon in midstream. That the French opera house was not only completed with magnificence but became a model for opera house design for at least three decades after its completion, and remains to this day a favorite venue for performers and public alike, attests to the value of the enterprise.

Rising construction and operating costs, dropping attendance, and aging music audiences are among the problems facing sites for sound, as is the preference of young audiences for alternatives to formal music venues. Yet some of the challenges embody positive considerations that have helped spur, in the first decade of the twenty-first century, an explosion of new buildings for the performing arts that equals the museum boom of the 1980s.

The Pompidou (Renzo Piano and Richard Rogers, 1977), the Menil (Renzo Piano Building Workshop, 1987), the Guggenheim Bilbao, and the 21st Century Museum of Contemporary Art, Kanazawa (SAANA, 2004), each introduced hitherto unforeseen ways of visiting art. In the same way that every decade has witnessed the birth of at least one museum that has influenced public perceptions of the plastic arts, so too will the new crop of music venues deliver enduring benchmarks for the performing arts. Current alterations in programming and presentation are comparable to what happened to art when, in the latter third of the twentieth century, it tumbled from its pedestal thanks in part to the changed circumstances in which it was experienced.

Forward-looking architects and acousticians are responding to ingenious new ways of performing music, and to accommodating new kinds of music, with equally inventive theater designs.

Zaha Hadid's Guangzhou Opera House, inaugurated in 2010, and two years later, the finishing of Coop Himmelb(l)au's International Conference Center, Grand Theater, and Opera House in Dalian could be the beginning of a move toward more imaginative architecture for this kind of facility in China. Likewise, Hadid's Heydar Aliyev Cultural Centre (2012) introduces a dramatically new kind of architecture to Azerbaijan. In South America, Europe, and East Asia other noteworthy complexes are underway or planned. For some of these—Christian de Portzamparc's Cidade da Música in Rio de Janeiro (2012), Herzog & de Meuron's Elbe Philharmonie in Hamburg (2014–15), Ateliers Jean Nouvel's Paris Philharmonie (2014), and Eisenman Architects' Ciudad de la Cultura in Santiago de Compostela (completion date not determined)—construction has been hampered by technical difficulties and political complications. Yet each enterprise exhibits a remarkably imaginative response to a demanding context. Toyo Ito's Taichung Metropolitan Opera House in Taichung City, Taiwan, is slated to open in 2013, and OMA/AMO hopes to launch its Taipei Performing Arts Center by 2015.

The dramatic facades of the Dalian and Baku complexes conceal rather predictable interiors, noteworthy for their colossal scale rather than for design innovation. Conversely, the exterior of the Taichung City opera house harks back to progressive trends of the 1960s, and it is the building's three theaters and circulation spaces that appear to be audaciously experimental. The Taipei project is a creative continuation of Rem Koolhaas's ongoing search for new solutions to the problem of theater design. Several recent concepts are driven by an equally iconoclastic approach.

Dalian International Conference Center, Grand Theater, and Opera House (Dalian Guoji Huiyi Zhongxin)

Dalian, China
Coop Himmelb(l)au, 2012
1,266,374 square feet
Cost undisclosed (estimated at $209.5 million)

Located at the tip of northern China's Liaodong Peninsula, the port city of Dalian, controlled by Russia's czarist government in the late nineteenth century, was conceived at that time as an alternative to Vladivostok. In the mid-1990s, Mayor Bo Xilai began to revive the city on a modern European model, turning it into an active port and popular resort. The new conference center, sited on the waterfront beside the Yellow Sea, marks the confluence of Dalian's two major urban axes, thereby aspiring to be a focal point in the spirit of Paris's Palais Garnier. Its spectacular site, jutting into the water, is reminiscent of that of the Sydney Opera House.

Reflecting the water, the soft curves of Coop Himmelb(l)au's giant clam shape undulate gently, as if "generated by the forces of the sea," in the words of principal Wolf Prix.[2] Like Jean Nouvel's KKL in Lucerne, and a number of subsequent structures, the building combines conference and performance spaces; in Dalian these facilities are situated almost fifty feet above a ground-level public zone containing shopping and exhibition areas. Thirty-six meeting rooms of varying sizes arranged around the largest conference space at the core of the main level add up to what Prix calls "a small city."[3] Some of these rooms pierce the facades, creating a faceted body. A 140-car parking garage, truck delivery, and waste disposal occupy the basement, where an adjacent 3,000-car underground garage is also planned.

The backstage wall of the 1,600-seat Grand Theater has steel doors that can open to the 2,500-seat conference hall behind it, making an enormous arena for popular events. The theater is a multiuse hall where amplification can be added to the natural acoustics required by classical music, including opera. Various surface treatments in the theater, notably large featherlike extensions on areas of the plastered balcony balustrades, are meant to enhance the acoustics. (The acoustics consultant is Eckard Mommertz of Müller-BBM.)

The conference center bears an interesting relationship to the Shenzhen Museum of Contemporary Art and Planning Exhibition (2013), another Coop Himmelb(l)au project under construction in China. Both have curved roofs consisting of multiple

Exterior view, rendering.

Shenzhen Museum of
Contemporary Art and
Planning Exhibition,
Shenzhen, China, 2013.
Rendering.

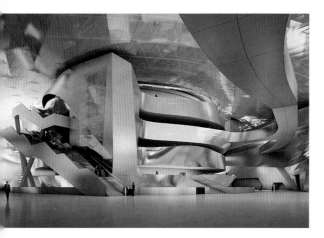

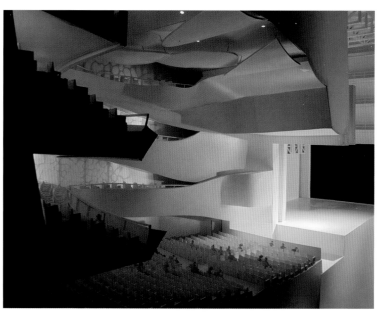

layers of metal and glass; in each case, these are covered by steel gridwork—in Shenzhen, a more solid punched-metal screen controls light; in Dalian, sun-reflecting louvers also control air flow. The systems provide daylight (supplemented by electric light) in the circulation areas of the two structures, an energy-saving feature; both projects also use solar and geothermal energy sources.

Like the other architects who were introduced to the New York public in Philip Johnson and Mark Wigley's "Deconstructivist Architecture" exhibition at MoMA in 1988—Frank Gehry, Daniel Libeskind, Rem Koolhaas, Peter Eisenman, Zaha Hadid, and Bernard Tschumi, all of whom have designed important music venues—Prix has enjoyed a phenomenal career in the intervening years. His Central Los Angeles Area High School #9 (2008) is a recent achievement in the United States, and in Europe, the BMW Center in Munich (2007) typifies his showy style. In the course of a visit to the BMW Center, the current mayor of Dalian, Li Wancai, was so entranced by its gigantic spaces and seductive automobile displays that he suggested to his city's Urban Planning Bureau that Coop Himmelb(l)au be invited to participate in the competition for the conference center.[4] Without reproducing his Munich work, the architect hopes to achieve equally dramatic effects with the Dalian project.

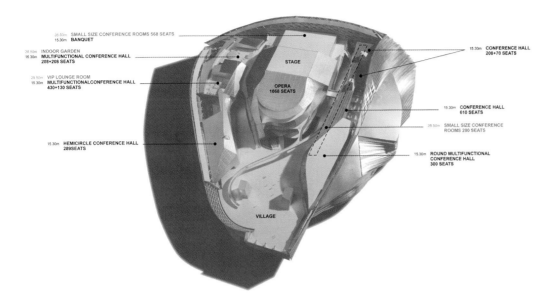

SMALL SIZE CONFERENCE ROOMS 568 SEATS
BANQUET

INDOOR GARDEN
MULTIFUNCTIONAL CONFERENCE HALL
208+208 SEATS

VIP LOUNGE ROOM
MULTIFUNCTIONALCONFERENCE HALL
430+130 SEATS

HEMICIRCLE CONFERENCE HALL
289SEATS

STAGE

OPERA
1668 SEATS

VILLAGE

CONFERENCE HALL
208+70 SEATS

CONFERENCE HALL
610 SEATS

SMALL SIZE CONFERENCE
ROOMS 280 SEATS

ROUND MULTIFUNCTIONAL
CONFERENCE HALL
300 SEATS

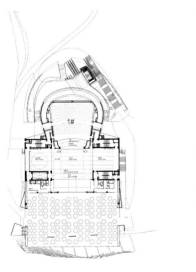

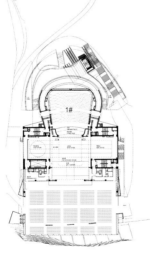

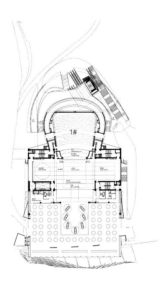

BANQUET AREA 1800 SEATS   CONFERENCE 2500 SEATS   SHOWS - CENTER STAGE

Heydar Aliyev Cultural Centre

Baku, Azerbaijan
Zaha Hadid Architects, 2012
1,095,777 square feet
Cost undisclosed

Of Zaha Hadid's various commissions for arts buildings, including those in Amman, Abu Dhabi, Dubai, Bonn, Rabat, and Chengdu, the nearest to completion is the Heydar Aliyev Cultural Centre in Baku, the capital of Azerbaijan, on the western coast of the Caspian Sea. Zaha Hadid Architects has connected the site's two levels, which are separated by a sixty-six-foot drop, so that they read as the surface of an inflected landscape, which will be used for outdoor activities and performances. A majestic stairway leading to the cultural center traverses this landscape. Bold curvilinear forms and their free flow from an exterior plaza into the lobby relate

Exterior view, rendering.

Concert hall, rendering.

to this architect's recent designs for other public buildings. The enormous folds of the facade and lobby precede a series of lower, rippling volumes; earth mounds in the plaza where the center is located also radiate away from the building like waves.

The project is located on the boulevard that links the old middle of Baku, now a city of nearly two million inhabitants, to the international airport. A road leading from this major thoroughfare to the cultural center bifurcates, with one spur going to the main entrance at the higher end of the site and the other to the 1,500-vehicle underground parking area. Achieving a striking contrast with the rigid, monumental architecture erected when this oil-rich nation was a constituent member state of the USSR (from 1936 to 1991), and visible from around the city, Hadid's building is the outstanding component of a developing neighborhood.

The different internal programs—conference hall, eight-story library, and museum—are expressed in the exterior massing; each has its own entrance and can operate independently. Mega-corridors, which integrate ramps, staircases, and escalators, connect the facilities, their design harmonizing with the rounded shapes of

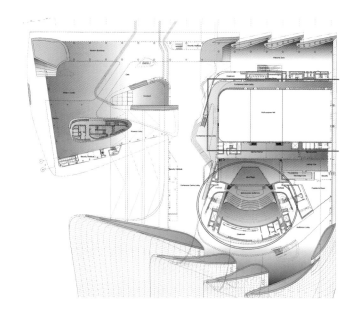

the facades. Communal zones will contain cafeterias, bars, restaurants, and meeting rooms, among other services.

The rectangular 1,200-seat conference hall can be divided into three separate spaces of different sizes. Of a piece with the center's general aesthetic, the auditorium interior is curvilinear and continuous, with finishes in solid oak.[5] It has been designed, with a hydraulic orchestra pit, for both natural and assisted acoustics to carefully tailor the sound (the acoustical consultant is MEZZO Studio). Again, this is a center that will host large conventions as well as musical performances. In fact, Azerbaijan has a century-long tradition of national classical music that draws on Iranian-Arab-Turkish systems of *maqam* (*mugham* in Azerbaijani): combining traditional modal and improvisational techniques with Western-style orchestral composition. The method of improvisation is sometimes compared to jazz.

Two systems respond to the city's high wind loads and location within a seismic zone. A space frame with a single movement joint allows construction of the free-form structure. Self-supporting glass-reinforced fiber-concrete and fiber-plastic panels are used for the envelope, making columns unnecessary. These planar, single-, and double-curved panels are extruded or molded. Whether the planned smoothness of the building's surfaces can be achieved will depend in large part on how skillfully the seams (incorporating expansion joints) between the cladding's disparate elements can be executed.

Heydar Aliyev (1923–2003), after whom the center is named, was a former Soviet police general and head of the KGB in Azerbaijan who ruled the impoverished eight million inhabitants of his native land for thirty years, first as its Communist leader, then, after independence, as its elected president. His son, Ilham, has succeeded him, and the center appears to be part of an ongoing continuation of the father's cult.

Cidade da Música

Rio de Janeiro, Brazil
Christian de Portzamparc, 2012
1 million square feet (approximate)
$304.6 million (approximate)

Christian de Portzamparc's cultural complex is located in Barra da Tijuca, a new, fast-growing town on Rio de Janeiro's southern outskirts, for which Lúcio Costa, best known for his plan (in collaboration with Oscar Niemeyer, 1956) for Brazil's capital, Brasília, had developed an extensive layout in 1968. Costa envisaged integrating residential towers with parks and nature reserves. That plan was abandoned when speculators gained control of the project, producing a dense mass of high-priced residential towers and commercial facilities with neither public space nor public transportation. All that remains of Costa's scheme are two regional highways. The Cidade da Música stands at their intersection, making a stunning finale for one. Portzamparc lifted the building off the ground on thirty-three-foot pilotis, helping

Exterior view; section.

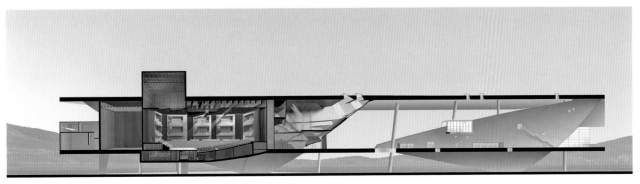

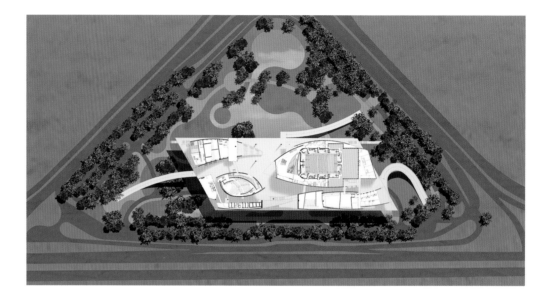

Site plan.

visitors distance themselves from the highways. But even the park around the Cidade, planned by the landscape architect Fernando Chacel, appears insufficient to bolster the complex from the onslaught of traffic amid which it is isolated.

Gently inclined ramps, stairways, and elevators lead up to a vast roofed, open terrace on which sit the closed volumes of three theaters. Characterized by plays of daylight between the complex shapes of these volumes and by expansive views of the Siera Atlantica mountain range, the space is called by Portzamparc, in the Brazilian tradition, a "veranda." In fact, the entire structure is informed by a vocabulary of abstract, thin concrete forms sandwiched between horizontal planes, which typifies the work of Niemeyer (his 1960 Planalto Palace in Brasília, for example). Portzamparc had tried out a similar, but less attractive, combination of conic forms within horizontal planes for his extension of the Palais des Congrès at the Porte Maillot in Paris (1999).

In addition to the Cité de la Musique in Paris (1985; see Chapter 3), the architect and his acoustician for music-related projects, Albert Yaying Xu, are responsible for the Philharmonie Luxembourg (2005). There, the main 1,500-seat concert hall is a traditional shoebox shape; eight towers, each with twenty-eight seats, are placed at different angles around the hall. In Tijuca, for the principal 1,800-seat Philharmonic Hall, the architect has elaborated on this design concept. Here there are again eight towers, four of which can be moved on air cushions to make a proscenium stage connected with the side wings. This reconfiguration, and the possibility of having an orchestra pit and a stage tower, transforms the auditorium into a 1,300-seat opera house. Additionally, there will be a 500-seat chamber music hall, a 180-seat hall for electronic and acoustic music, three movie houses (one with 300, two with 150 seats), a mediathèque, a restaurant, and stores.

César Maia, the mayor of Rio, invited Portzamparc to design and build the Cidade in 2002, when it was expected to cost $123.8 million. By the time construction began two years later, the program had expanded, increasing costs, which rose still further when local engineers had difficulties with the demanding construction.

The project has also been handicapped by two government-ordered work stoppages. Between 2005 and 2007, construction was halted when the city decided to mobilize its resources for the sports facilities it was building in Rio for the 2014 World Cup soccer tournament and the 2016 Olympic and Paralympic Games, the latter to take place in part in Barra da Tijuca. In 2009 activity was again halted, this time by a newly elected mayor, Eduardo Paes, who, in anticipation of public criticism of the escalating expenditures, by then over $290 million, ordered a public inquiry.

At the end of 2010, construction slowly resumed. By early 2011 only the reinforced-concrete exterior was complete: giant, gracefully curved planes that billow out between two long (115 feet), thin (five feet) horizontal slabs. The unfinished building, which was to have been the first cultural and social amenity in the area, stands forlorn in an area severed from its surroundings by chaotic traffic. As has occurred at other arts complexes, the people in charge in Rio prioritized architecture over programming and only now are seeking a private partner with whom to share funding of the Cidade's estimated annual operating costs of $7 million.

Taichung Metropolitan Opera House

Taichung City, Taiwan
Toyo Ito, 2013
372,446 square feet
$145 million (estimate)

Toyo Ito's work is well represented in Europe and East Asia, and his Sendai Mediatheque in Japan (2000) has received international attention. Ito's first United States commission was for a building in California: a 2006 replacement for the University of California, Berkeley's Art Museum and Pacific Film Archive (Mario Ciampi, 1970). Unfortunately, Ito's stunning design was unrealized due to a lack of funding.

In size and structural complexity, the Taichung Metropolitan Opera House is second only to the Sendai building. The two are radically dissimilar in appearance, however. Contrasting with the transparent modernist glass box of the Mediatheque are the unexpected organic forms of the opera house in Taichung, a city of over a

Exterior view, rendering.

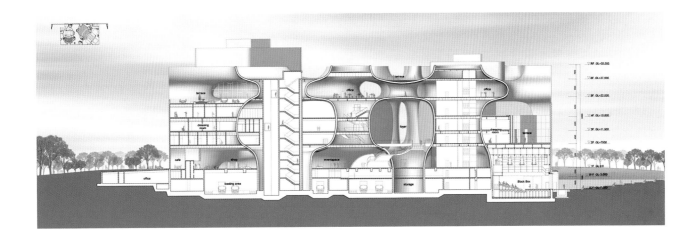

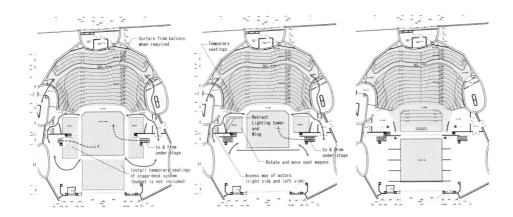

million inhabitants in north-central Taiwan. The roots of these shapes may be found in Japan's Metabolist movement of the early 1960s, often associated with the "plug-in" designs of Archigram in London.

Ito first explored the ideas proposed in the Taichung Opera House in a competition organized by Gerard Mortier in 2004 for the Forum of Music, Dance and Visual Culture in Ghent, which was not built. For Taichung's program, Ito modified his earlier scheme; of particular note, he made the performance spaces more flexible. (Toshiko Fukuchi of Nagata Acoustics, Tokyo, serves as consultant.) Included are a 2,011-seat grand theater for Western, Chinese, and Taiwanese opera, classical ballet, and contemporary dance; an 800-seat theater for drama, Chinese and Taiwanese opera, dance, and ballet; a 200-seat black-box experimental theater; and shops, restaurants, and cafés, all set within a park. The three theaters can be reconfigured,

Section; plans showing three configurations: center stage, thrust stage, proscenium.

Grand theater,
rendering; site plan.

the larger one with the help of a mechanized orchestra lift and seating wagons. These wagons are also present in the medium theater, where lighting towers can be retracted. A manual stage-deck system provides flexibility in the black-box theater.

Within an enveloping rectangular box of eight stories—six above ground and two below—the Opera House consists of a honeycomb of horizontal and vertical spaces based on a generative grid in which the two-dimensional plans for each level are developed into areas with a unifying, three-dimensional curved surface. Looking something like superimposed egg cartons, and called "sound caves" by the architect, they form a continuum of foyers, theaters, rehearsal rooms, workshops, and restaurants with views from one to another.[6]

The homogeneity and flexibility of the theaters recall designs by the visionary architect and artist Frederick Kiesler for a cavelike Universal Theater, which evolved between the 1930s and the 1960s and which would have destroyed the traditional divisions between floor, wall, and ceiling and permitted various stage and seating arrangements. Ito shares with Kiesler an interest in architecture without beginning or end, a forerunner of the current "architecture of continuity."

The building is constructed by means of a truss wall system almost fifteen inches thick. Its painted reinforced-concrete and glass facades afford views in all directions, with entranceways at every side. Surfaces are concrete sprayed onto expanded metal mesh, a material combination commonly used for the construction of tunnels. This open structure is intended to actively engage its surroundings, encouraging encounters between high and popular art, artists and visitors, stage and auditorium, interior and exterior. The rooftop terrace's network of water and greenery, similar to that of the city, further connects the opera house with the park.

Paris Philharmonie

Paris, France
Ateliers Jean Nouvel, 2014
215,278 square feet
$559 million

In October 2010, a surprise announcement that work on Jean Nouvel's Paris Philharmonie at the Parc de la Villette had been stopped for eight months, with no date set for its resumption, raised concerns about what the French press described as "the biggest cultural building site of the last five years."[7] The uncertainty was particularly painful to Pierre Boulez and other prominent French musicians who wished to bring an additional venue for classical music to this working-class district of the French capital. As far back as 1984, Christian de Portzamparc had included in his competition-winning scheme for the nearby Cité de la Musique a large auditorium for symphonic music. But it remained only a concept.

Awarding of the Philharmonie project to Ateliers Jean Nouvel (working with both Marshall Day and Yasuhisa Toyota) in late 2007 revived hope. A 2,400-seat main auditorium, with unusually ample rehearsal facilities and an educational wing, could be Paris's major concert hall, bringing to life a new topology with its surround seating in continuous, cantilevered balconies reached by narrow bridges. With the audience standing, the way it does for London's Proms, as many as three thousand could be accommodated.

The hall, designed for the natural acoustics suited to classical music, will also have amplification for jazz and world music. Shortly after the project was announced, Exterior view, rendering.

Concert hall, rendering; plan; sections.

Laurent Bayle, director of the Philharmonie Association, spoke enthusiastically of the advantages for productions that would be afforded by seven hydraulic platforms within the audience space to accommodate off-stage players.[8] Additionally, the hall's walls will provide cycloramic lighting that can be modified according to the performance.

Squeezed between the Cité de la Musique and a portion of the eight-lane Boulevard Périphérique, the site is inhospitable, recalling the similar, even more difficult, problem Nouvel had in Copenhagen. However, like that for the Danish Radio Concert Hall, the architect's scheme—here, a dramatically sloping landscape of gigantic aluminum slabs—conveys its own strong image. Masking the highway at the northeast, the Philharmonie's shimmering, stacked planes will face the park (mirroring the nearest of the bright red follies built in 1992 by Bernard Tschumi) and reach out to the adjacent Cité de la Musique; its uppermost areas will overlook Paris.

The walkway of the Philharmonie's ramped roof, which contains the main entrance, recalls the Oslo Opera House by Snøhetta (see Chapter 3), but its multiple layers are more intricate than those in Norway. The building's oblique forms also hark back to the collaborative work of the architect Claude Parent and the theorist Paul Virilio. In fact, Nouvel says he conceived the design as a homage to Parent, in whose office he worked as a young assistant from 1967 to 1970, when the older practitioner was developing his theory of intentionally disorienting oblique architecture.[9]

Repeating what has become a predictable pattern, the Philharmonie's first cost estimate, in 2007, of $154 million had by the summer of 2009 nearly quadrupled, to $559 million ($419 million of which was allocated to construction). Forty-five percent of the initial tab was to be picked up by the state, forty-five percent by the municipal government, and ten percent by the Ile-de-France region. By December 2010 the federal government's failure to honor its commitment, which had brought construction to a halt, was overruled by President Nicolas Sarkozy, and work resumed early in 2011.

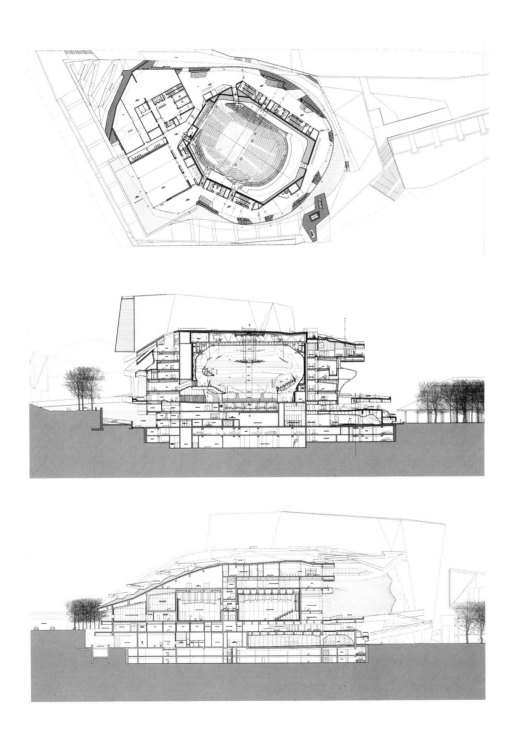

Elbe Philharmonie

Hamburg, Germany
Herzog & de Meuron, 2014–15
1.3 million square feet (approximate)
More than $620.5 million

When Herzog & de Meuron broke ground in April 2007 in Hamburg for the stunning Elbe Philharmonie—part of a structure that includes a luxury hotel, apartments, offices, and abundant commercial amenities—completion of the project was announced for September 2010, and the cost was estimated at $266.5 million. By May 2010, inauguration of the mixed-use enterprise had been postponed until 2013, with a new cost estimate of more than $620.5 million.[10] And in late 2011, the completion date was again put back.

The architectural scheme crowns a former cocoa bean warehouse, the Kaispeicher A. (Werner Kallmorgen, 1963), with a breathtaking giant prism, called the Crystal. The curved panels of the glass walls—twelve stories on one side, eighteen on the other—develop a quilted effect that recalls the firm's Prada Aoyama Epicenter in Tokyo (2003), and the wavy roof culminates in a dramatic profile reminiscent of snow-capped mountain peaks. The glass's deformation allows openings in the facade that contribute to a rich variety of reflections.

Unfortunately, the sturdy appearance of the brick warehouse proved deceptive. Once additions had been made to the original program, including a 518-car garage, 620 additional reinforced-concrete piles were needed to support the structure. Pierre de Meuron attributes the rise in costs to technical flaws related to acoustics and to the lack of coordination between the architects, the builder, and the city, a recurring problem for public projects.[11]

Herzog & de Meuron has long been known for visually bold, structurally rigorous accomplishments, among them the de Young Museum in San Francisco and a freestanding addition to the Walker Art Center in Minneapolis (Edward Larrabee Barnes, 1971), both completed in 2005. It was, however, the extensive television coverage of the architects' striking Beijing National Stadium, constructed in consultation with the Chinese artist Ai Wei Wei for the 2008 Olympics, that gave the architects worldwide recognition.

Mindful of the increased contact between public and athletes achieved by the stadium's design, Herzog & de Meuron took it, rather than Scharoun's Berlin Philharmonie, as a model for the main 2,150-seat concert hall of the Crystal. Still, the Elbe's surround seating and tentlike roofline clearly acknowledge the architects'

admiration for Scharoun's example. As Jacques Herzog says, "The Berlin hall is perfect; it can't be improved."[12]

Because the Crystal's form is dictated by the truncated triangle of the original building, the plan of the auditorium was compressed in order to fit. Consequently, it is higher than the Berlin Philharmonie, rising in places to more than ninety feet, and therefore distancing from the musicians those sitting in the upper level (but bringing nearer those in lower levels). The Elbe Philharmonie goes beyond the older hall's segmental vineyard seating, stacking seats in continuous, torqued balconies with limited overhangs, and thus resulting in a more fluid space.

Acoustically isolated from the rest of the building, the 12,500-ton auditorium rests on 362 giant spring assemblies. (There is also a 550-seat shoebox chamber music hall and, in the older part of the building, a 170-seat theater.) The so-called "white skin" of the main hall's walls was developed with acoustics in mind (the consultant is Yasuhisa Toyota): it is comprised of dense gypsum fiberboard panels hollowed out in an egg-crate pattern. Harmonizing with rough oak floors (like those of the firm's renovated Tate Modern, 2000) and tweed-covered seats, the panels are intended to retain the natural, gritty feel of the site.

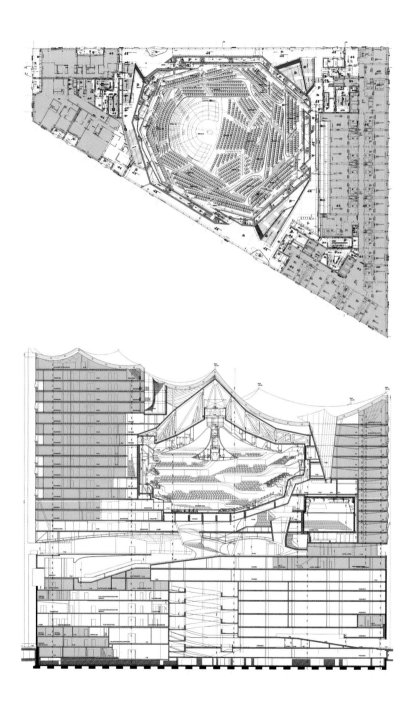

Small niches throughout the main hall allow the spatial placement of musicians in the auditorium, in addition to onstage. However, Christoph Lieben-Seutter, director of the Elbe Philharmonie, notes that these niches will be used when works call for them, mostly in performances of the Romantic pieces preferred by Hamburg's audiences.[13]

The Philharmonie is part of the $13-billion-plus HafenCity plan, among the most ambitious urban renewal projects in Europe. Following a master plan approved in 2000, construction targets the former port, south of Hamburg's city center, which will be expanded by forty percent when completed by 2020 or 2025. In addition to large international corporations in their own high-profile buildings, more than two hundred smaller businesses will be based in the area, which will also contain generous residential facilities. About 1,500 residents have moved into the renovated harbor district, and approximately 6,000 people commute to work there.

Already a major tourist attraction, the Elbe Philharmonie is the star of the new precinct. A 269-foot escalator will rise diagonally from the wharf to a semi-outdoor terrace between the brick structure and the Crystal, which commands panoramic views of the harbor and the city. The combination of new steel-and-glass buildings and handsome nineteenth-century brick warehouses, which echo Hamburg's historic facades, make HafenCity an unusually attractive area on the banks of the Elbe River. If the Philharmonie lives up to its promise, the cost increases and completion delays will soon be forgotten.

Plan; section; concert hall, rendering.

Ciudad de la Cultura

Santiago de Compostela, Spain
Eisenman Architects with Andrés Perea Ortega and Antonio Maroño
Components: Galician Library, National Archive, spring 2011;
Museum of Galician History, Central Services and Administration, fall 2011;
International Art Center, preliminary construction begun in 2011;
Performing Arts Theater, scheduled to begin construction 2014
Total of four buildings: 1 million square feet
Performing Arts Theater: 370,000 square feet
$600 million (approximate for all components)

Planning for this vast cultural campus, larger than the Getty Center in Los Angeles, began in 1999 with an initial estimate of $146.4 million for the six buildings (reduced from eight) of Peter Eisenman's competition-winning project.[14] Eisenman, like Daniel Libeskind (see Chapter 3), began his career as a teacher and theoretician.

Exterior view, 2011.

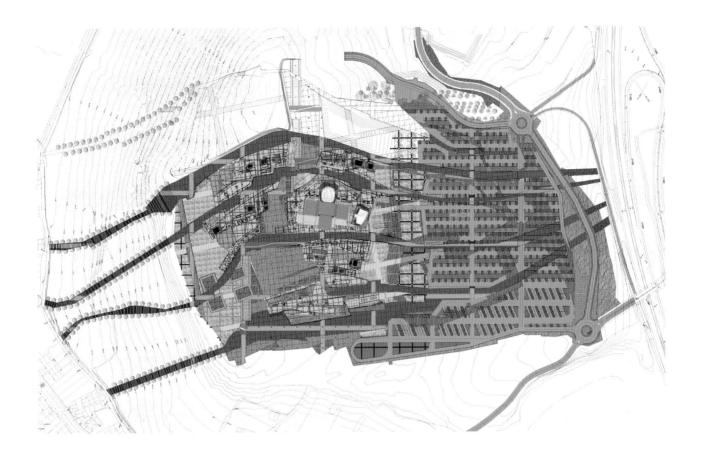

As the founder and first executive director of the nonprofit Institute for Architecture and Urban Studies in Manhattan (1967–84), he exerted a strong progressive influence on the profession. The Wexner Center for the Arts at Ohio State University (1989), designed with Richard Trott, was Eisenman's first noteworthy public commission; the Santiago project is his largest to date.

Construction started in 2001 and was expected to take three years. By early 2011, the white facades and rose, brown, and off-white quartzite roofs of four buildings (of which only two had been completed) were visible: the Galician Library, National Archive, Museum of Galician History, and Central Services and Administration crown Monte Gaiás on the outskirts of Santiago de Compostela. When finished (the Central Services and Administration interiors, the International Art Center, and the Performing Arts Theater), the Ciudad de la Cultura will have cost nearly $600 million to build, and it could require more than $80 million a year to run—a whopping thirty percent of the budget for culture, communications, and tourism of Galicia, a poor agricultural area. Like Portzamparc's Cidade da Música, the undertaking was perhaps overambitious at the outset and became more so as it developed, sharing many of the Rio center's problems of construction, cost overruns, and political swings.

The ninety-thousand-inhabitant city of Santiago has been a pilgrimage site since the ninth century, currently attracting twelve million of the faithful each year to its famous Romanesque (with Baroque exterior) cathedral, reputedly the burial place of St. James the Apostle. In the 1980s the Socialist mayor, Xerardo Estévez, initiated a modernization plan that was to balance historic preservation related to the city's religious role with the implementation of cultural infrastructure. The most interesting

Entrance; library.

result of this effort is Álvaro Siza's handsome Centro Gallego de Arte Contemporanea (1994) in Porto, completed at a time of regional competition with Bilbao's planned Guggenheim Museum and Valencia's Ciudad de las Artes y de las Sciencias.

In 1999, a new Conservative local government, under president Manuel Fraga Iribarne, continued, and apparently attempted to trump, Estévez's initiative with the Ciudad de la Cultura, commissioning Eisenman Architects. (Another conspicuous architectural project, a telecommunications tower to be designed by Norman Foster, was not undertaken.) Even before design development, the size of Peter Eisenman's design tripled: among other increases, the Performing Arts Theater—incorporating spaces with 1,560 and 450 seats—nearly doubled to 99,000 square feet for the main theater and stage and 370,000 square feet for the whole building (it will be the largest of the six structures), and the library's capacity was expanded from 250,000 to one million books.[15]

With the Socialists' return to power in 2005, work was stalled in order to investigate a new appraisal of approximately $365 million.[16] It resumed three years later with additional deviations from the program. The slowdown and change orders incurred further cost increases, and along the way, the Eisenman team was replaced by Spanish practitioners (for working drawings and construction supervision).

The complex was conceived as seismic folds within the dour, granitic Galician landscape, its buildings accessed by small passageways scaled like the old city's medieval streets and linked by large tunnels. In a favorite procedure of Eisenman's, the design evolved from distorting the medieval city's plan, which in this case is overlaid with a grid and twisted into a fractured geometry.[17] The scallop-shell shape and ridge

Performing Arts Theater,
interior view, rendering;
interior perspective.

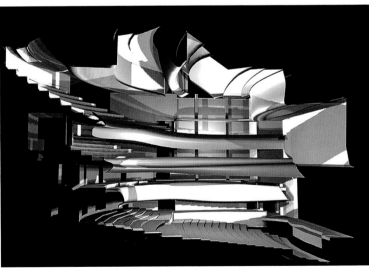

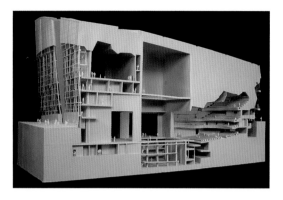

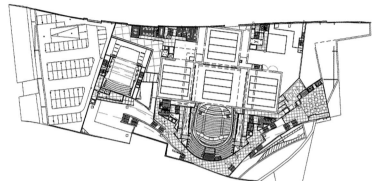

of the center's overall plan relate to the emblem of St. James and the pilgrimage route. Because the architect envisaged the project as a landscape, considerable effort has been made to harmonize sheathing and paving materials with the surrounding environment (landscaping is planned by Laurie D. Olin of OLIN).

The three facilities that had opened by the end of 2011 are impressive: seen from afar, they appear to be a series of giant stone waves freezing movement in time (as does the work of such architects as Gehry and Hadid). The reinforced-concrete structures are covered in finely worked, twenty-inch-square blocks of quartzite mounted on a steel armature. Facades are different on every side, and large expanses of glass—mainly on the south of the archive building and the north of the library—give a sense of transparency, allowing interesting plays of light in the soaring, sculpturally contoured public areas.

Interiors are punctuated by large and medium concrete columns placed on two grids; together with the library's plentiful trusses, required by ninety-eight-foot ceilings, the effect is somewhat overwrought. The spaces of the two buildings completed early in 2011 are remarkably generous, but only about a quarter of the areas is in use, their immensity disproportionate to the needs of this relatively small community.

Judgment of the Ciudad de la Cultura must, however, await its completion; in 2011, the complex had a gap at the center (like a mouth missing some teeth) into which the two final structures will fit. The horseshoe interior of the larger theater in the Performing Arts Center is yet another pairing of conservative interior and progressive exterior. The articulated ceiling and balcony fronts are intended to make dynamic surfaces.[18] With yet another switch in local politics having taken place in 2009, the remaining two structures may not be up and running for some time.

Taipei Performing Arts Center

Taipei City, Taiwan
OMA/AMO (Office of Metropolitan Architecture and in-house
design and research studio) with Artech Architects, 2015
430,556 square feet
$134.5 million

Rem Koolhaas, in his usual provocative mode, dismisses recently built performance
spaces as "versions of a more or less identical combination of large auditorium,
medium theatre, and black box ... contemporary forms disguising conservative typologies."[19] Conversely, he calls his own performance spaces "machines" or "engines."[20]
In the Taipei Performing Arts Center, three theaters will function independently or
in various combinations. The technical apparatuses are consolidated into a single
cube, and the stages can be joined when needed.

Model.

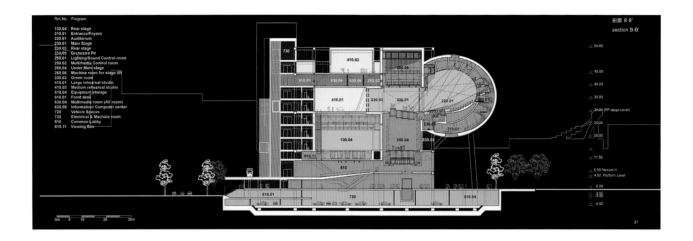

Koolhaas points to OMA's Congrexpo in Lille, France (1994), as an early example of the scheme that his firm continued to elaborate for the Ghent Forum competition (2004) and for the Wyly Theatre in Dallas (2009). The spaces of the three designs were to be reconfigurable; in the case of Congrexpo, a five-thousand-seat hall for popular concerts and three conference auditoriums can be connected to make an enormous area. These ideas are taken even further in the Taipei Performing Arts Center.

The center has no single front facade. Rather, it has four building faces with differently shaped theaters projecting from three of them. The technical facilities for the theaters will be consolidated in a central cube, a seismically isolated steel structure sheathed in corrugated glass. Koolhaas claims the concept was spurred by his adverse reaction to Paul Andreu's 2007 National Center for the Performing Arts in Beijing (see Chapter 4), where, he observes, "everything is repeated three times with no additional effect."[21] Reminiscent of the Congrexpo spaces, but considerably smaller, those in Taipei can be used individually or together. The acoustician for the project, Renz L. C. J. Van Luxemburg, explains that the theaters will have natural acoustics when functioning independently and amplification when combined.[22] The building has extensive refreshment areas, numerous rehearsal and changing rooms, offices, storage space, and below-grade parking facilities for 200 cars and 350 scooters.

A slightly asymmetric rectangle with a progressive rake encloses the 1,500-seat Grand Theater for Western and Chinese contemporary opera, musicals, dance, ballet, and drama. It projects from the south side of the cube on slanted supports. Entrances to the theater are at the back and from the sides, the latter connecting with a cross aisle that is level with the stage, thereby providing an open control position for the electronic sound system in the main audience area. The orchestra pit, stage, and some parterre seating are automatically adjustable.

A sphere, poised on a V-shaped support at the cube's east side, contains the eight-hundred-seat Proscenium Playhouse for drama, traditional Chinese opera, dance, musicals, and puppet shows (in a small theater below). Its inclined parterre seating is embedded in stacked balconies arranged in a modified horseshoe. Without a permanent frame, the proscenium can be adjusted in size. Both the orchestra pit and the stage are similarly flexible.

The eight-hundred-seat, flat-floor Multiform Theater can be transformed from end stage to thrust to arena. A balcony at the back has narrow extensions along the side walls. The theater's rectangular form, supported on columns, extends from the north side of the cube on axis with the Grand Theater, making it possible for the two theaters and the intermediary technical area to connect as one giant space, even joining with the grand theater's fly tower.

The Taipei Performing Arts Center is located within a short walk of the MRT railway station, in the midst of one of the largest and most popular of the many night market districts in this densely populated city of 2.62 million. The hope is that crowds will be channeled to the center's common lobby from which some escalators

Plans.

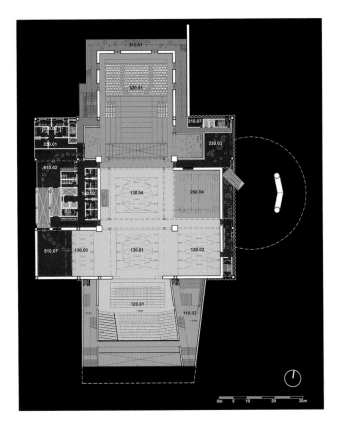

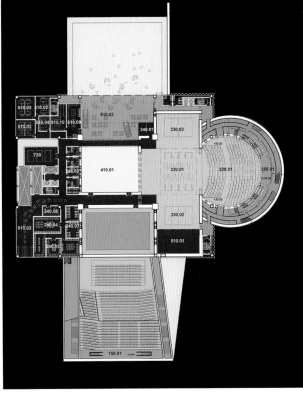

Grand Theater, rendering; Proscenium Playhouse, rendering; Multiform Theater, rendering; Super Theater (Multiform and Grand Theaters combined), rendering.

will rise to the theaters while others will take nonticketed visitors to backstage views and to a rooftop café. Below grade an enormous food court stretches the width of the building in what Koolhaas calls "a gradual transition from the basic Asian condition to a more refined theater condition." The architect describes the station, the night market, and the cube as "forming a cluster of three urban densities, connected by a continuous human flow."[23] In this city of vitally active theater, where stage directors are cult heroes, OMA/AMO's Performing Arts Center is slated to be a major attraction as well as an important influence on the evolution of theater design.

HOW WILL THE MOST RECENT VENUES AND THOSE STILL UNDER CONSTRUCTION measure up acoustically? Since even the most conventionally shaped auditoriums can be disappointing in this respect, aural satisfaction becomes even more elusive when wedded to experimental architecture.

My own listening experiences favor Yaying Xu's conciliatory description of acoustics as an "exact science, but with an undeniable element of creative subjectivity."[24] In a recent interview, Yasuhisa Toyota articulated the parameters by which he and most experts in this field judge acoustics: "If an acoustician can ensure that the audience is no longer aware of how far away they are positioned from the music, he has done a good job."[25] Poor acoustics "make the sound appear to be too far away"; thus Toyota's prime aim is to achieve what he calls "acoustic intimacy."

Eckard Mommertz, acoustical consultant for the Grand Theater in Dalian, emphasizes the difficulty in harmonizing, according to him, "quiet acoustic engineering" with a flamboyant design.[26] Even before provision was made for acoustical surfaces, in response to his advisors, Wolf Prix narrowed the auditorium's width and protected it with two concrete walls, thereby providing aural insulation within the building's steel structure (metal is more susceptible to sound transmission than concrete). Furthermore, floors rest on resilient foundations so that they avoid transmitting the structure's vibrations. Only usage will reveal whether these measures are enough to deal with the demands of a complex theater expected to serve a multitude of uses.

Another enigma is whether the acoustics of Zaha Hadid's Baku cultural center auditorium will match those of her highly praised Guangzhou Opera House. Before its completion, Guangzhou's slight asymmetry raised doubts about its sound; it would be ironic if the strictly symmetrical hall in Baku, where Hadid is working with a different consultant (MEZZO Studio instead of Marshall Day Acoustics), turned out to be less satisfactory.

After a trial run of the orchestra in the main hall of Rio's Cidade da Música in 2008, at a time when eighty percent of the interior work was still incomplete, the artistic director and principal conductor Roberto Minczuk said, "Even unfinished, the big auditorium can be compared with the best in the world . . . the acoustics are fabulous with very rich harmonious sounds, enhancing sonority and the pitch of each instrument without losing clarity."[27] But in 2010, it was reported that the rush to finish construction caused problems with the concert hall's acoustics that have made it necessary to demolish and rebuild parts of it.[28]

The experimental architecture of the Taichung Opera House presents a similar acoustic problem to that of the Dalian theater. Because the concave shapes of all three of the Taichung performance spaces are exactly what acousticians try to avoid, the outcome here is especially unpredictable. Toyo Ito and Toshiko Fukuchi have collaborated closely since the design competition in 2003. After tests of the Grand Theater's wall and ceiling reflections via a computer-simulation system and a one-tenth-scale acoustical model, the architect adjusted interior shapes and in places has added convex surfaces together with rough, uneven materials. Additionally, a doughnut-shaped sound reflector will hang under the opera house ceiling.[29] The partially outdoor "sound cave" circulation spaces raise an additional set of questions about the quality of sound in them.

The competition for the Paris Philharmonie stipulated that each proposal include an acoustician. Two separate acoustical consultants ended up working on the project—Marshall Day and Nagata. The record of conflicting approaches among members of the profession raises the question of whether teamwork in this sensitive area can be effective.

For Hamburg's Elbe Philharmonie, Toyota's engineering is optimized for traditional music; Christoph Lieben-Seutter points out that neither the hall's acoustics nor its sight lines favor the rock concerts (amplified by the firm dUCKS scéno) that will constitute twenty-five percent of the programming.[30] It would appear that the space's unusual height could make it difficult to achieve the desired acoustic (or visual) intimacy for those sitting in the hall's upper regions, while the plan's compression might reinforce intimacy in the lower areas.

There are unresolved issues regarding the acoustics at the Ciudad de la Cultura Performing Arts Theater, for which Mark Holden, the acoustician who worked with Eisenman Architects, recalls a program that specified opera and dance, not concerts. "This is why there is such a big backstage and elaborate orchestra pit," Holden explains. "It was not to be a full music theater."[31] As for most of the theaters discussed here, amplification was to allow performances in a range of genres. Now, with no director in place, it is anyone's guess how the two spaces for music will be used, and to what effect acoustically.

Possible acoustical failings presented by combining different spaces with OMA's three theaters at the Performing Arts Center in Taipei appear to be resolved by the use of amplification when the auditoriums are enlarged. There remains, however, among other questions, the concern that because of their juxtaposition, noise generated by preparations in one theater might be audible in another. The gamble undertaken by Rem Koolhaas and Joshua Prince-Ramus on a vertical, modular theater paid off acoustically for the Wyly in Dallas; the equally innovative nature of the plug-in theaters in Taipei has its own incertitudes.

ACOUSTICAL QUALITY IS NOT ALONE AMONG OUTSTANDING QUESTIONS FOR opera houses and concert halls of the future. There is also uncertainty about the funding of programs and maintenance and the related issue of making tickets more affordable. Projections of aging and declining audiences and the dearth of public arts education are matters of concern. The preference of younger people for alternatives to formal theaters appears to challenge the latter, as does the staggering success of high-definition broadcasts and streaming to home computers and portable devices (although producers are convinced that broadcasts and streaming will build support for the art form). And the effectiveness of palatial new theaters in improving under-privileged urban neighborhoods has yet to be proven conclusively.

Despite these negative considerations, building and planning multimillion-dollar arts centers on campuses around the United States has not been hindered by current cutbacks in state support for higher education.[32] That many of these com-plexes are largely state-financed indicates that they are considered essential, as are the hundreds of private venues being built in this country.[33] Among the opera houses and concert halls that have opened, or are about to open, in North America within the past decade are the Kimmel Center for the Performing Arts in Philadelphia (2001) and Manhattan's Jazz at Lincoln Center (2004)—the Rose Theater, the Allen Room, and Dizzy's Club Coca-Cola—by Rafael Viñoly Architects; the Adrienne Arsht Center for the Performing Arts of Miami-Dade County (2006) and the Renée and Henry Segerstrom Concert Hall in Costa Mesa, California (2006), by Pelli Clarke Pelli Architects; the Laura Turner Concert Hall at the Schermerhorn Symphony Center in Nashville (2006) by David M. Schwarz Architects; the Kauffman Center for the Performing Arts in Kansas City, Missouri (2011), by Safdie Architects; the Montreal Concert Hall (2011) by Diamond+Schmit; Dr. Phillips Center for the Performing Arts in Orlando (projected 2012) by Barton Myers Associates; and the Bing Concert Hall at Stanford University (projected 2012) by Ennead Architects.

Several of these encountered major funding problems. The Kimmel, the Arsht, and the Segerstrom produced enormous budget deficits. In the case of the Segerstrom, the shortfall was made even more onerous by increased interest rates and the issu-ance of additional bonds.[34] The Arsht, which contains an opera house, concert hall, and theater, was dealt another setback when the resident Florida Philharmonic, the state's major symphony orchestra, declared bankruptcy and was dissolved in the spring of 2003, shortly after construction of the center began. The Florida Philhar-monic is one of several important orchestras that have in recent decades sustained serious financial difficulties. Whereas the hundred-million-dollar renovation of Lincoln Center's David H. Koch Theater only increased the problems of New York City Opera, the Nashville Symphony, which resorted to Chapter 11 in 1988, was saved in large part by its move to the new Schermerhorn Symphony Center (coupled with a

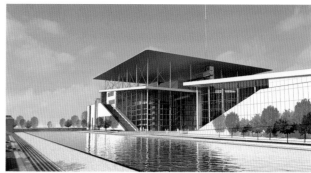

Harpa Concert Hall
and Conference Center,
Reykjavik, Iceland, 2011.
Exterior and interior
renderings.

Stavros Niarchos
Foundation Cultural
Center, Athens, Greece,
2015. Exterior and
interior renderings.

restructuring of its operations). Ticket sales role from $2.6 million in the 2005–6 season to more than $7 million in the 2006–7 season.[35]

In the United States, where cultural institutions are mostly private endeavors, the hurdles of fundraising for day-to-day operations can prove even more difficult than those of fundraising for construction. Often, board members who are remarkably generous toward high-profile architecture to which their names are attached can be indifferent to the cost of running these institutions. Putting enduring architecture on the map is more attractive to many donors than enabling an ephemeral performance. That ongoing support is essential to the survival of organizations that are rarely financially self-sufficient was understood by Arthur and Judy Zankel. In addition to underwriting construction, the donors of the Zankel at Carnegie Hall provided funds to subsidize programs and operations.

Michael M. Kaiser, president of the John F. Kennedy Center for the Performing Arts since 2001 and a guru of arts management, maintains that to gain consistent support creative management must develop a wide circle of donors by alternating sure-fire hits with venturesome repertory—the latter often bypassed because of increased rehearsal time and the uncertainty of critical reaction.[36]

Matching the profusion of new music venues in North America are those in the rest of the world. Among the most ambitious are the Harpa Concert Hall and Conference Center in Reykjavik (2011) by Henning Larsen Architects with Batteríio (incorporating a facade devised with the artist Olafur Eliasson); the Helsinki Music Center (2011) by LPR-arkkitehdit Oy; and the Stavros Niarchos Foundation Cultural Center in Athens and a concert hall in Bologna (both 2015), by Renzo Piano

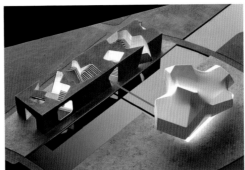

Building Workshop. In 2009 Christian de Portzamparc was awarded the project for CasaArt, a performing arts center in Casablanca (the architect's birthplace); in 2010 Herzog & de Meuron was commissioned to design the Cultural Complex Luz (including music and dance schools plus three performance halls) for a depressed area of São Paulo; also in 2010, the Piano office won the competition for the Thăng Long Opera House in Hanoi. One of several additional performing arts centers being planned in China is a Grand Theater by Steven Holl at the Normal University in Hangzhou, which supplements Carlos Ott's Hangzhou Grand Theater of 2004.

Many of the same financial problems found in the United States exist for these foreign projects, where local or national governments that have consistently paid for construction are drastically reducing their contributions to cultural activities, in all likelihood on a permanent basis. This has already happened in France, Great Britain, Spain, and Italy, where the major burden of underwriting now lies with private philanthropy. Given the complaints of Chinese musicians about insufficient operational funding, even that prosperous nation may have to consider non-government largesse. Introducing individual and corporate fundraising, however, will not be easy for countries with little or no tradition of such financial support.

Considering the mixed results, the use of high-profile cultural facilities as catalysts for poor or newly planned urban neighborhoods needs some rethinking. Not all new buildings have exerted as successful a draw as the Oslo Opera House and the Muziekgebouw in Amsterdam. The Arsht Center has not substantially improved the security problems in that area of downtown Miami, and despite its relatively good financial standing, the New Jersey Performing Arts Center in Newark (1997) by

Cultural Complex Luz, São Paulo, Brazil, 2016. Exterior and interior renderings.

Grand Theater, Normal University, Hangzhou, China, 2015. Exterior and interior renderings.

Barton Myers failed to revive that city's business district. Even the universal acclaim for Frank Gehry's Walt Disney Concert Hall in Los Angeles (2003), and the stimulating effect the hall has had on the area surrounding it, have not yet reversed the stalled plan to overhaul the Bunker Hill neighborhood. Such a time lag may plague projects with yet greater challenges, like the Jean Nouvel–designed Danish Radio Concert Hall, which is intended to humanize a bleak new suburb of Copenhagen.

As worrisome as the financial struggles of these costly buildings are the discouraging statistics about the audiences for classical music events that are expected to fill them. For the United States, the League of American Orchestras posted a thirteen percent decline in attendance between 2002 and 2008, and it predicts an additional fourteen percent drop by 2018.[37] The National Endowment for the Arts (NEA) saw an even greater decline (twenty percent) and found that audiences for classical music, ballet, non-musical theater, and jazz were aging faster than the general adult population (from a median age of forty in 1982 to forty-nine in 2008, and for opera from forty-three to forty-eight).[38] French attendance seems to be fairly stable, even rising in Paris in the 2008–9 season (although dropping slightly nationwide).[39]

Germany, until now a bastion of music boasting the densest concentration of opera houses and concert halls in the world, has already seen a slight decrease in the number of its symphony orchestras and, with the exception of Berlin, Dresden, and Munich, a falling off in opera attendance. Because of the low birth rate there in the late 1960s, the total number of concertgoers is also expected to diminish appreciably by 2020 or 2025. In addition, music audiences in Germany tend to be a decade older than audiences in the United States.[40]

Kaiser, for one, blames the absence of arts education in the United States since the late 1970s for depriving those entering their forties of an arts background.[41] Typically, people who were exposed to the arts as children put aside these activities as teenagers and young adults and returned to them in their forties and fifties. Absent this sequence, there is not only less revenue from ticket sales but a smaller donor base.

Among the many paradoxes of today's music world is the disconnect between the increasing number of artists and nonprofit arts organizations since the 1960s (with the help of the NEA and the states' arts agencies) and the neglect by education policy-makers of instruction in the knowledge and skills needed to appreciate offerings in the arts. Aggressively countering this dereliction are numerous private initiatives to supplement arts education.[42] Major institutions across the United States extend arts instruction to classrooms and to individuals, usually free of charge: most orchestras, opera companies, and chamber music and jazz organizations have substantial education departments. The Lincoln Center Institute (founded in 1975) and Carnegie Hall's Weill Music Institute (founded in 2000), both located in Manhattan, have reached millions of students worldwide. In 2007 Carnegie Hall, the Juilliard School,

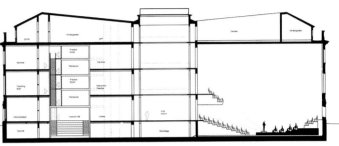

West-Eastern Divan
Institute, Berlin, Germany,
2013. West facade;
proposed section.

and the Weill together created the Academy for beginning professional musicians; Carnegie Hall is adding a sixty-thousand-square-foot educational space for the Weill, scheduled to open in 2014, to accommodate this program. Carnegie Hall's educational and community measures, in which 170,000 partake annually, are augmented by a web site for educators, and a new achievement program developed in collaboration with the Royal Conservatory in Toronto is addressed to nonprofessional as well as aspiring professional musicians.[43]

Similar efforts contributing to middle- and high-school musical-instrument training include Chicago's Civic Orchestra (a training orchestra run by the Chicago Symphony since 1919), the Chicago Young Musicians Initiative (begun in 2009), the San Francisco Symphony Youth Orchestra (begun in 1981), and Opus, also in San Francisco (developed as a pilot program in 2001 and launched as the Institute Training and Support Program in 2006). A precursor to Miami's New World Symphony, the National Orchestral Association, which has supported musicians, composers, and conductors since 1930, evolved in the 1990s into a multidimensional sponsor of performance groups and educational programs.

A comparable problem plagues Germany: prior to the austerity measures of 2010, music education in that country (geared to appreciation rather than performance) was compulsory from grades one to ten. Deep cuts in funding have made music a nonessential subject, and it is being dropped from curriculums. Following the conviction that "fewer and fewer people will go to concerts unless music becomes part of general education,"[44] Daniel Barenboim is expanding the West-Eastern Divan Orchestra (WEDO, which he founded in 1999 with the late Palestinian scholar Edward Said) into a permanent, year-round institution dedicated to humanities education through music. It is scheduled to start operations in 2013 with about forty students (in their late teens and early twenties), twenty to thirty graduate students, and a few senior fellows in a historic warehouse of the State Opera House (Richard Paulick, 1955) in the heart of Berlin. The interior will be renovated or entirely rebuilt and will include a seven-hundred-seat concert hall; the budget for the renovation is between

Modular Music Factory, prototype, São Paulo, Brazil, 2012. Rendering; section.

$18 and $30 million. The work will be carried out by the local firm HG Merz, with Frank Gehry contributing design concepts.

Among WEDO's programs is the Music Kindergarten, a privately run, government-subsidized undertaking inaugurated in 2005 that implements the concept of inter-disciplinary education for preschool children. (The Niarchos Foundation Cultural Center is one of several institutions that include a small theater for children's opera.) Barenboim says the Aspen Music Festival and Institute as it was founded in 1949 was an inspiration in its linking of music and the humanities in an interdisciplinary context that stresses international understanding.[45] Aspen's approach arose from the devastations of World War II, WEDO's from the divisions in the Middle East.

Just as the West-Eastern Divan Orchestra combines a political purpose—encour-aging peace in the Middle East—with an educational mission, the National System of Children and Youth Orchestras of Venezuela, or El Sistema, applies music to socioeconomic aims, but with a less personalized and more wide-ranging structure and operational mode. Begun in 1975 by the Venezuelan conductor, pianist, and politician José Antonio Abreu with the Simón Bolívar Youth Orchestra of Venezuela, the organization now boasts 180 orchestras throughout the country, in which 290,000 children participate. The program, which reaches out to the most disadvan-taged urban areas, serves as a model for music, culture, and education in dozens of countries, especially in the Americas. Among its distinguished graduates is Gustavo Dudamel, music director of the Los Angeles Philharmonic, which will guide a national teaching program to be called Take a Stand, based on El Sistema.[46]

In collaboration with El Sistema, the Centros de Acción Social por la Música in

São Paulo, Brazil, began construction in 2011 (for completion in late 2012) of a replicable prototype for a music school in the Paraisópolis favela. The building, designed by Alfredo Brillembourg and Hubert Klumpner of Urban-Think Tank and estimated to cost $5 million, found its site when a massive mudslide swept away a maze of shacks, creating the Grotão, a flood bowl. As future iterations may be, the building is erected on top of a sports field (the only open terrain in Brazilian slums) next to which the flood bowl is landscaped. Envisioned as a music factory, the structure is organized vertically so that it can be inserted into tight sites in densely populated poor neighborhoods. The prototype incorporates a concert hall (with acoustics by Müller-BBM) and accommodates eight hundred to a thousand students a day.

Young people are by no means turning away from classical music. On the contrary, the enormous number of small, non-traditional listening places for such music that have come into being worldwide since the 1980s testifies to an ongoing, keen interest. The popularity of alternative spaces recalls the movement of art exhibitions in the 1960s and 1970s away from museums and established art galleries to renovated lofts. Today there is a related trend for classical, jazz, and world music—and for experimental new music.

New music has, since the late 1960s, resembled modern fine arts—painting, watercolor, and photography—in its combination of styles that were once strictly segregated. And like the visual arts, new music currently has no dominant approach. Compare the earlier generation of new music composers, like Morton Feldman, John Cage, Olivier Messiaen, Luciano Berio, Gyorgy Ligeti, or older composers still living, like Pauline Oliveros, Philip Glass, Steve Reich, Arvo Pärt, with midcareer composers, like John Luther Adams, Jacob TV, Mark O'Connor, Kaja Saariaho, or with the extremely polystylistic younger generation, like Nico Muhly, Timothy Andres, Bora Yoon: the variety and combinations of styles these artists present are dizzying and becoming ever more so. Additionally, numerous groups that specialize in performing new music—Eighth Blackbird, Bang on a Can All-Stars, ICE (International Contemporary Ensemble), Alarm Will Sound, Argento Chamber Ensemble, Ensemble Modern, the Kronos Quartet—encourage and sustain the introduction of increasingly divergent ideas into the classical world: rock, world music (either particular ethnic traditions or, as with Yo-Yo Ma's Silk Road Initiative, reimagined cultural borrowings from along an ancient trade route), bluegrass, jazz, improv, chance, games and algorithms, feedback, electronics, distortion, sampling, choreography, noise, birdsong, commercial jingles, multiphonics, speech patterns, manipulated speech, tonal, microtonal, serial, spectral, minimal, maximal, modal, mashup, and the list goes on. Because of a curiosity about wildly diverse musical genres and techniques, it is no wonder that people who frequent the spaces where these groups perform have been called cultural "omnivores."[47] If contemporary audiences were like those in

Le Poisson Rouge, New York, New York, 2008.

the eighteenth century, who preferred to hear the latest compositions of their contemporaries, like Mozart and Haydn, new music would be the focus of mainstream concert-going audiences.

The attraction of listening outdoors, in cabaret style, and in other similarly unconventional conditions demonstrates that for today's emerging audiences intensified intimacy and a user-determined experience compete in importance with quality acoustics. There is obviously a willingness to trade perfection of the latter for a taste of the former. Roulette (1978), established in a West Broadway residential loft and relocated in 2011 to a 650-seat Art Deco theater in Brooklyn, was one such space. Others include the Knitting Factory (1987), occupying formerly dilapidated offices near the Bowery in Manhattan; Knitting Factory Entertainment now offers music and art in cafés around the country—so far in Brooklyn, North Hollywood, Boise, Spokane, and Reno. Galapagos Art Space in Brooklyn (1995) resides in an old mayonnaise factory. Le Poisson Rouge, located in a basement formerly occupied by the Village Gate in an Ernest Flagg building (1896) in Manhattan, is a particularly successful example: a multimedia cabaret founded in 2008 by classical musicians David Handler and Justin Kantor, it offers popular and art programs in a 250-seat room, capable of seating and stage rearrangements. On the Boards in Seattle (1978, in its present location since 1998), Yellow Lounge in Hamburg (2001), Berlin, and Dresden (both 2004), Classical Underground in Los Angeles (2008), the Church of Beethoven in Albuquerque (2008), and King's Head Theatre & Pub in London (established in the 1970s and in 2010 made into a full-time opera house) are among the numerous unconventional places that have cropped up—with acoustics of variable quality.

Departure from the hidebound behavior expected in formal theaters is perhaps another way to allow the public to escape the strictures of conventional venues. Eighteenth-century opera and concert patrons shouted, clapped, and stamped their approval, stimulating the players to take risks and keep the feedback coming.[48] Clearly, younger—and possibly some older—audiences are resisting top-down cultural experiences in favor of events they can appropriate as part of their own milieu: these participants download, sample, and interact with live performances, as they do with popular concerts and performance art.

Performers seem to share the public's ease in these informal settings. When the Chamber Music Society of Lincoln Center repeated, in spring 2011, an earlier concert later in the evening with wine and candlelit tables in the society's small Rose Studio, one critic found the interpretations more organic, the instrumental textures more clearly defined, and the playing more energetic.[49] In 1975, for the first of the Rug Concerts in Avery Fisher Hall (see Chapter 2), David Hamilton had noted a similar phenomenon in the *New York Times*: "There was a greater warmth and urgency in [Boulez's] music making" than in the regular season.[50]

Performances in non-traditional spaces are facilitated by significantly reduced remuneration for the musicians that in turn allows lower prices of admission, and the remarkable turnouts for less expensive, or free, events suggest that the desire is there, but that ticket prices for most live productions are too steep. An enthusiastic audience of twenty-three thousand, approximately thirty percent between the ages of twenty-five and forty-four, attended a no-cost 2008 simulcast of *Lucia di Lammermoor* at San Francisco's AT&T Park. And in 2010 the Metropolitan Opera's free outdoor concerts in four boroughs attracted an audience of nearly eighty thousand. In a more popular vein, but including classic favorites, was "Three Tenors in Concert" (1990s to early 2000s), for which the opera singers Plácido Domingo, José Carreras, and Luciano Pavarotti drew enormous crowds to performances in stadiums and other large spaces.

Early in 2010, Michael Kaiser pointed out that for the cost of a pair of top-priced tickets to an opera, a spectator could buy a computer and watch the performance—and many subsequent ones—on YouTube or medici.tv for free.[51] As if to confirm this fact, and the outreach of widened technology, live streaming from the Sydney Opera House of its YouTube Symphony Orchestra 2011 performance was the largest such event on record: 30.7 million hits, with another 2.8 million viewers on mobile devices (also a record). The concert, which was heard in thirty-three countries on five continents, was called by Michael Tilson Thomas, who conducted, "a collaboration of classical music and technology."[52]

And there is the enthusiastic response to live, high-definition broadcasts in movie theaters, often with repeated showings. Pioneered in 2006 by the Metropolitan Opera in New York, the series sold 2.4 million movie theater tickets in the 2009-10 season; in 2010-11, twelve of the Met's operas went out to about 1,600 theaters in fifty-four countries.[53] The Royal Opera in London and La Scala in Milan are among the houses that offered live feeds in 2010-11, as did the Los Angeles and Vienna Philharmonics and the Philadelphia Orchestra. Even the strictly controlled performance environment of Bayreuth's Festspielhaus was opened to thousands in 2011 with a live transmission of *Lohengrin*. To date, only the Met has turned a profit on these broadcasts.

Like the elusive area of acoustics, there are no clear-cut answers to the complex questions that inhabit today's music scene. Conceivably, even the dynamic of positives and negatives may contribute to the field's current liveliness. One such ambiguity is the statement by Rem Koolhaas, a skillful theater architect, that just as art is best exhibited in nonmuseum spaces, the most exciting performances are seen outside the confines of traditional theaters.[54] An example of this was the 2006 presentation of Bernd Alois Zimmermann's 1964 opera *Die Soldaten* in a renovated Rhine steelworks in Bochum, Germany. In 2008 the production traveled to the mammoth Wade Thompson Drill Hall, reminiscent of a nineteenth-century train shed, at Manhattan's Park Avenue Armory (1881). David Pountney, the opera's director, says, "It is exciting to go outside the box."[55] Yet he cautions that such elaborate productions in nontheater stagings are extremely costly. (Acoustic adjustments are among the many special provisions needed for each new production in the Drill Hall.) More recently, Emilia and Ilya Kabakov's production of Olivier Messiaen's opera *St. François d'Assise* was performed in the ten-thousand-seat Madrid Arena (4,300 seats were made available) to ecstatic reviews. The production, commissioned by Gerard Mortier, the artistic director of the Teatro Real, is reported to have cost about $2.8 million.[56]

Die Soldaten was one of the first music events in a cultural series organized by the Park Avenue Armory Conservancy, a group that is programming the Drill Hall with art that cannot be mounted in formal museums or performance spaces. In order to achieve this in an aesthetically satisfying and cost-efficient manner, the conservancy included the Drill Hall as part of a $200 million renovation of the building it commissioned from Herzog & de Meuron. Finished in early 2011, work in the eighty-foot-high barrel-vaulted ceiling of the fifty-five-thousand-square-foot hall (costing $35 million) included acoustical consulting by Paul Scarbrough of Akustiks.

Seven hundred and thirty seats can be arranged at three sides of the flat stage area, or the space can be left free of seats as it was for a 2011 performance of John

Luther Adams's *Inuksuit*, a seventy-minute work originally meant to be played out-doors and inspired by the geography, ecology, and native culture of Alaska, since 1978 the composer's home state (this was its first performance indoors). The 1,300 attendees paid a relatively inexpensive thirty dollars each to wander among the seventy-six musicians (largely percussionists) scattered throughout the hall and its soaring entrance corridors. Unusually diverse listeners, both age-wise and socially, sat and lay on the floor or circulated, greeting friends and exchanging thoughts quietly, absorbed by this magic moment. Subtly audible ambient street sounds were piped in through loudspeakers to bring the outdoors in.

It would be a stretch to equate this event with a U2 or Lady Gaga concert in a sports stadium with upward of sixty thousand fans dancing in place and waving their cell-phone cameras to the music. Nevertheless, interaction between the public and the performers at the armory was tangible—people quickened their walking pace with the music's crescendos then closed their eyes and became contemplative during softer passages. Ever since the experimental theaters of the 1950s, directors have been mingling audiences and action, as did the Happenings of the 1960s; less common for musical theater, this phenomenon occurred at the armory, where the audience pro-vided the spectacle. The concert offered an experience that was at once profoundly communal and—because musicians and audience members were all in motion, with no two listeners hearing exactly the same thing—individual to an extreme.

Giving each spectator a particular audiovisual perception, and at the same time a communal experience, is a design concept for opera devised by the architect Valerio Maria Ferrari, whose small staff is divided between Paris and Milan. In addi-tion to his other projects, Ferrari has worked closely with singers and choruses (in collaboration with the director Piero Faggioni, he has staged operas at many leading houses), which has brought him to what he calls "a new way of listening": the Visual Music Facilities Theatre (VMFT).[57] Originating as a student project in the early 1990s,

Inuksuit performance, Park Avenue Armory, New York, New York, 2011.

Lady Gaga performance, Chicago, Illinois, 2010.

Visual Music Facilities
Theatre, rendering of
conceptual scheme, 2010.

inspired by Frederick Kiesler's "Space Stage" (1924), and resembling experimental
configurations such as the double-eight arrangement at MUMUTH in Graz (see
Chapter 3), the 2010 scheme consists of a snail-shaped stage with six hundred pivot-
ing seats interspersed within the spiral and eight hundred seats in a surrounding
balcony, the latter emulating a courtyard theater like the Globe.[58] The orchestra and
conductor are positioned at the center, and LED light effects could transform the
entire hall into a huge screen. The VMFT proposes a visual experience based on projections, three-dimensional technology, and interactivity that Ferrari considers more
radical than a modular typology, which re-creates known formations. The concept
has garnered approval from several acousticians.[59]

Gerard Mortier, on the other hand, rejects the scheme, which he feels would be
limited to works composed for the space. It is precisely this specificity that appeals
to Nigel Redden, director of the Lincoln Center Festival and general director of the
Spoleto Festival USA, who sees it as potentially enhancing any production that would
be prepared for it.[60]

The new buildings, projects, and concepts discussed here epitomize the diverse
environments in which music can now be enjoyed. The pleasure of these spaces
reinforces the joy of the sound. An increasing number of sites for sound have the
ability to escape convention inexpensively with unique events that can be reproduced

elsewhere. Furthermore, the greatly expanded capacity of new opera houses and concert halls to accommodate a wide range of acoustic and staging requirements allows more leeway in programming different genres, making these spaces more viable financially and perhaps even permitting more affordable tickets.

Rem Koolhaas and Alfredo Brillembourg refer to the "theater machine," a phrase Erwin Piscator used early in the twentieth century. Striving for such a "machine" since 2006 is a group of leading musicians led by Pierre Boulez; Boulez, Claudio Abbado, Riccardo Chailly, Esa-Pekka Salonen, and others wish to construct a new experimental opera and music theater in Lucerne. Introducing the project online, the description asks: "What would Claudio Monteverdi's music sound like without the acoustical experience he gained at Saint Mark's Basilica?"[61] This reference to the reciprocity between composer and cathedral takes as a given the historical relationship between music and the space for which it is composed: music does serve the space, as indeed the space serves the music. The two are inseparable.

Introduction

1. See Martha Feldman, "Festivity and Time," in *Opera and Sovereignty: Transforming Myths in Eighteenth-Century Italy* (Chicago: University of Chicago Press, 2007).
2. Frank Gehry, conversation with the author, July 18, 2008.
3. Throughout this book I use the word *music* to refer to classical music, including contemporary classical. For some interesting aspects of defining "classical music," see Greg Sandow, "Rebirth: The Future of Classical Music," ch. 4: "What Classical Music is," www.gregsandow.com/BookBlog/Ch4Riff.pdf.
4. Here, too, there is disagreement among acousticians: J. Christopher Jaffe insists that "the acoustics of the audience area has been solved through accepted scientific procedures" (email to the author, October 20, 2010); Yasuhisa Toyota says he owes his success to the fact that "I'm praying a lot" (conversation with the author, March 4, 2010).
5. For instance, additional demands are posed by the increasing interest in usually quieter, more delicate original and archaic musical instruments. These are used in spaces quite different from those where they were heard in their own time.
6. Laura Moretti, "Architectural Spaces for Music: Jacopo Sansovino and Adrian Willaert at St. Mark's," *Early Music History* (Cambridge: Cambridge University Press, 2004), vol. 23, 153-83. See also Richard Taruskin, *The Oxford History of Western Music*, vol. 1, *Music from the Earliest Notations to the Sixteenth Century* (Oxford: Oxford University Press, 2005), 599-606.
7. Claudio Monteverdi, who became *maestro di cappella* in 1613, continued to break down static conceptions of concert music in ways that still inspire composers far beyond the walls of San Marco.
8. Leo Beranek, *Concert Halls and Opera Houses: Music, Acoustics, and Architecture* (1962; New York: Springer, 2004), 494, 593; different acousticians cited by Beranek offer varying reverberation times.
9. This opinion was expressed by Christopher Blair, a conductor and acoustician, and John F. Allen of High Performance Systems, a sound and loudspeaker designer, at a panel discussion, "Concert Hall Design: How Much Art? How Much Science?" produced and moderated by Allen, November 16, 2008, and transcribed in *Boston Audio Society Speaker* 31 no. 2 (June 2009): 24.
10. Esa-Pekka Salonen conducted his Piano Concerto at the Musikverein on December 12, 2010.
11. The distinction between multiuse (small theaters acoustically adjustable to limited, specific needs) and multipurpose (large theaters with an infinite number of uses for which acoustic variability is insufficient) is made by George C. Izenour, "Multiple Use: A Contemporary Definition," in his *Theater Design* (2nd ed., New Haven: Yale University Press, 1996), 307-9.

Chapter 1
The Past: A Historical Overview

1. See Richard Taruskin, *The Oxford History of Western Music*, vol. 1, *Music from the Earliest Notations to the Sixteenth Century* (Oxford: Oxford University Press, 2005), 823-26.
2. See Fritz Graf, "Religion and Drama," and Richard P. Martin, "Ancient Theatre and Performance Culture," in *The Cambridge Companion to Greek and Roman Theatre*, ed. Marianne McDonald and J. Michael Walton (Cambridge: Cambridge University Press, 2007), 55-71, 36-54.
3. George C. Izenour, *Theater Design* (New Haven: Yale University Press, 1977, 1996) is an invaluable source of information on the history of theater design and acoustics. I have based my overview of theaters in antiquity on this source, unless otherwise indicated.
4. Richard Beacham, "Playing Places: The Temporary and the Permanent," in McDonald and Walton, *Greek and Roman Theatre*, 202-26.
5. Beacham, "Playing Places."
6. Pliny the Elder, *The Natural History*, ed. John Bostock, H. Y. Riley, book 36, ch. 24.
7. David Whitwell, "Essays on the Origins of Western Music," www.whitwellessays.com/docs/DOC_478.doc.
8. Plutarch, *Life of Pericles* 13, trans. Bernadotte Perrim, Loeb Classical Library 65 (Cambridge, Mass.: Harvard University Press, 1916, 2001), 43.
9. George C. Izenour, *Roofed Theaters of Classical Antiquity* (New Haven: Yale University Press, 1992), 30-35, and Beacham, "Playing Places."
10. My description of the Odeum of Agrippa is based on Beacham, "Playing Places," and Izenour, *Roofed Theaters*, 84-89.
11. J. Christopher Jaffe, *The Acoustics of Performance Halls: Spaces for Music from Carnegie Hall to the Hollywood Bowl* (New York: W.W. Norton & Company, 2010), 12.
12. Izenour, *Roofed Theaters*, 63-64.
13. Fernando Rigon, *The Teatro Olimpico in Vicenza* (Milan: Electa, 1989), 48, 71 n. 23.
14. Izenour, *Theater Design*, 266.
15. See Simon Tidworth, "The Picture and the Frame," *Theatres: An Architectural and Cultural History* (New York: Praeger Publishers, 1973).
16. Michael Forsyth, *Buildings for Music: The Architect,*

the Musician, and the Listener from the Seventeenth Century to the Present Day (Cambridge, Mass.: MIT Press, 1985), 73–76.

17. Ellen Rosand, *Opera in Seventeenth-Century Venice* (Berkeley: University of California Press, 1991), 14. See also Martha Feldman, *Opera and Sovereignty: Transforming Myths in Eighteenth-Century Italy* (Chicago: University of Chicago Press, 2007), 145–46.

18. See Rosand, "*Da rappresentare in musica*: The Rise of Commercial Opera," in *Opera in Seventeenth-Century Venice*, 66–81, 179.

19. Feldman, *Opera and Sovereignty*, 35.

20. Forsyth, *Buildings for Music*, 71.

21. The historic opera houses mentioned here have all undergone extensive rebuilding and/or renovation.

22. All references to Richard Pilbrow are from his conversations with the author, May 2 and July 2, 2008.

23. Frederic Spotts, *Bayreuth: A History of the Wagner Festival* (New Haven: Yale University Press, 1994), 51.

24. Forsyth, *Buildings for Music*, 181.

25. Izenour, *Roofed Theaters*, 181–86.

26. J. Christopher Jaffe, email to the author, October 10, 2010.

27. Built between 1778 and 1784 (and destroyed by fire in 1958), Claude-Nicolas Ledoux's theater at Besançon also abolished boxes and anticipated Wagner's screened, sunken pit.

28. For a convincing examination of the centrality of the *Gesamtkunstwerk* concept to modernism in the visual arts and architecture, see Juliet Koss, *Modernism after Wagner* (Minneapolis: University of Minnesota Press, 2010).

29. Spotts, *Bayreuth*, 13.

30. Daniel Barenboim, conversation with the author, August 10, 2010.

31. Forsyth, *Buildings for Music*, 27–43.

32. Forsyth, *Buildings for Music*, 55.

33. H. Bagenal and Godwin Bursar, "The Leipzig Tradition in Concert Hall Design," *Journal of the Royal Institute of British Architects*, September 21, 1929, 756–63.

34. Jaffe, *Acoustics of Performance Halls*, 14–15.

35. Forsyth, *Buildings for Music*, 152.

36. Roger Williams, *The Royal Albert Hall: A Victorian Masterpiece for the Twenty-first Century* (London: Fitzhardinge Press, 2003), 117.

37. The concert hall was restored between 1979 and 1984 after its destruction in World War II and renamed the Konzerthaus. The interior departs from the original.

38. Both the Musikverein and the Concertgebouw have full pipe organs, considered by many essential for a symphony hall. This instrument (together with a chorus behind the stage) also absorbs some of the sound from the high-energy brass instruments and timpani at the rear of ensembles and helps balance the sound generally.

39. Alex Ross, conversation with the author, February 13, 2008.

40. Tim Blanning, *The Triumph of Music: The Rise of Composers, Musicians and Their Art* (Cambridge, Mass.: Belknap Press of Harvard University, 2008), 96–97.

41. Lynne Conner, "In and Out of the Dark: A Theory about Audience Behavior from Sophocles to Spoken Word," in *Engaging Art: The Next Great Transformation of America's Cultural Life*, ed. Steven J. Tepper and Bill Ivey (New York: Routledge, 2008), 103–26.

42. Koss, *Modernism after Wagner*, 240.

43. David Robertson, conversation with the author, September 23, 2008.

44. Daniel Barenboim, conversation with the author, August 8, 2010.

45. Jaffe, *Acoustics of Performance Halls*, 89.

46. Blanning, *Triumph of Music*, 108.

47. Jaffe, *Acoustics of Performance Halls*, 93.

48. Apparently even performers' gender and body language influence the public's assessment of their playing; Tom Jacobs, "Visual Cues Impact Judgment of Piano Performances," Miller-McCune.com, August 29, 2011, http://www.miller-mccune.com/culture/visual-cues-impact-judgment-of-piano-performances-35579/. And Allan Kozinn said of the JACK Quartet's performance in an entirely dark hall that "It would be hard to think of a more involving way to hear a new work"; "A Musical Question: Does a Quartet Sound Different in the Dark?" *New York Times*, September 7, 2010, C3. For broader references to the effect of acoustical characteristics of built form, see Kenneth Frampton, *Studies in Tectonic Culture* (Cambridge, Mass.: MIT Press, 1995), 9.

49. Yasuhisa Toyota, conversation with the author, March 4, 2010.

50. Gerard Mortier, conversation with the author, October 8, 2009.

51. Among the many surround halls that followed Scharoun's are Town Hall Auditorium in Christchurch, New Zealand (Warren and Mahoney, 1972), the Sala Nezahualcoytl, Mexico City (Orso Nunez, Arcadio Artis, Manuel M. Ortiz, Arturo Trevino, and Roberto Ruiz, 1976), and Suntory Hall, Tokyo (Yasui Architects, 1986).

52. See Marc Treib, *Space Calculated in Seconds: The Philips Pavilion, Le Corbusier, Edgard Varèse* (Princeton, N.J.: Princeton University Press, 1996).

53. "Expo ABZ," *Architectural Design and Construction*, June 1970, 281.

54. Kathinka Pasveer, Stockhausen-Stiftung für Müsik, email to the author, July 20, 2011.

55. "Expo ABZ," 281.

56. Arata Isozaki, "From Panopticon to Archipelago," *GA Document* 57: 55-63.

57. Renzo Piano, conversation with the author, February 13, 2008.

58. Luigi Nono, "Possibilité et nécessité d'un nouveau Théatre Musical," *Ecrits* (Paris: Editions Bourgois, 1993).

59. Whitwell, "Essay Nr. 46: On Ancient Concert Halls," "Essays on the Origins."

60. Michael Tilson Thomas, "Mahler: Origins and Legacy," *Keeping Score* (DVD, San Francisco Symphony, 2011).

61. Merce Cunningham, conversation with the author, August 22, 2008.

62. Jeffrey T. Schnapp, "Border Crossings: Italian/ German Peregrinations of the *Theater of Totality*," *Critical Inquiry* 21 (autumn 1994): 80-123. In the 1960s, leading Americans also regarded the arts as an antidote to the alienation of modern life: Donna M. Binkiewicz, *Federalizing the Muse* (Chapel Hill: University of North Carolina Press, 2004), 53.

63. Schnapp, "Border Crossings."

64. Kenneth Frampton, *Modern Architecture: A Critical History* (New York: Oxford University Press, 1980), 139-40.

65. Yasuhisa Toyota, conversation with the author, October 19, 2007.

66. See Introduction, note 11.

67. Jaffe, *Acoustics of Performance Halls*, 101.

68. George R. Steel, conversation with the author, April 28, 2008. Steel's statement will be put to the test with his decision in 2011 to leave the Koch Theater and stage his productions in various spaces around the city. He said, "Now we have the right operas in the right spaces." James R. Oestreich, "Ushering City Opera into a New Era," *New York Times*, August 12, 2011, C7.

69. Ara Guzelimian, conversation with the author, January 25, 2008.

70. Boulez said the idea for *Répons* came to him during a visit to Frank Lloyd Wright's Guggenheim Museum in New York City, where visitors can simultaneously view what they have seen and what they will see, allowing past and present to interact. Jean-Pierre Derrien, interview with Pierre Boulez, quoted by Allen B. Ruch, http://www.themodernword.com/joyce/music/Boulez_repons.html.

71. See Maria Luisa Neri and Laura Thermes with Alfonso Giancotti and Carlo Serafini, *Maurizio Sacripanti: Maestro di Architettura 1916-1996* (Rome: Gangemi Editore, 1998).

72. Pierre Boulez, conversation with the author, February 26, 2008.

73. Laurent Bayle, conversation with the author, March 21, 2008.

74. Mark Swed, email to Raphael Mostel, June 6, 2011.

75. Albert Yaying Xu, email to the author, June 27, 2011.

76. Clive Gillinson, conversation with the author, February 28, 2008.

77. James Conlon, unpublished article, 2009.

78. Gerard Mortier, "From Basilica to Forum: Views with Regard to the Construction of a Performance Hall for the Future," 2004, unpublished essay.

79. Renée Fleming, conversation with the author, February 27, 2008.

80. Peter Sellars, conversation with the author, May 1, 2008.

Chapter 2
Lincoln Center: "From Behind Walls to the Street"

1. Unless otherwise indicated, my accounts of the origins of Lincoln Center, with the exception of the Metropolitan Opera, are based on the unpublished report to the Lincoln Center Development Project, Inc., by Deborah Fulton Rau, "The Evolution of Lincoln Center's Site Plan 1956-59: The Historical Development of the Buildings and Outdoor Spaces," July 2004.

2. Ada Louise Huxtable, then architecture critic of the *New York Times*, argued forcefully for a regenerative downtown location for the Kennedy Center in her article "From a Candy Box, a Tardy and Unpleasant Surprise," *New York Times*, August 22, 1965. Georgetown residents' objections to a subway connection had no effect on the WMTA decision against the station: see Zachary M. Schrag, *The Great Society Subway: A History of the Washington Metro* (Baltimore: Johns Hopkins University Press, 2006).

3. Lincoln Center originally came in at over $165 million—more than $1 billion in 2011 dollars. See Norval White and Elliot Willensky with Fran Leadon, *AIA Guide to New York City*, 15th ed. (New York: Oxford University Press, 2010), 354.

4. Introduction, "Lincoln Center for the Performing Arts: An Analysis of Existing Conditions, Historical Development, and Future Plans" (unpublished report prepared for the Lincoln Center Development Project, Inc.).

5. See Victoria Newhouse, *Wallace K. Harrison, Architect* (New York: Rizzoli, 1989), chapter 17.

6. Herman Krawitz, interview, January 29, 1991, "Lincoln Center for the Performing Arts, Inc., Oral History Project," 61. Subsequent references to this source are cited as "Oral History Project."

7. Edgar B. Young, interview, June 11, 1990, "Oral History Project," 72.

8. Those who perform in Alice Tully Hall, part of the Juilliard School Building, lease space from Lincoln Center for the Performing Arts, Inc., as do the New York Philharmonic Society at Avery Fisher Hall and the Metropolitan Opera Association at the Metropolitan Opera House; Lincoln Center for the Performing Arts, Inc., leases the Koch Theater from the city of New York.

9. Pierre Boulez and Rolf Liebermann, "Opera Houses?—Blow Them Up!" *Opera*, June 1968, 440-50.

10. Young, interview, 94.

11. Max Abramovitz, interview, July 10, 1990, "Oral History Project," 111.

12. Ada Louise Huxtable, conversation with the author, February 26, 2009.

13. The firm of Diller + Scofidio was chosen after Frank Gehry's 2001 proposal for a glass dome over Lincoln Center's main plaza was rejected.

14. David Rockwell, conversation with the author, September 1, 2009.

15. Ada Louise Huxtable, "Juilliard's New Building: Esthetic Reality," *New York Times*, October 8, 1969, 59.

16. This quote and other information and quotes from Elizabeth Diller are taken from conversations with the author, February 26 and April 18, 2009.

17. Huxtable, conversation, February 26, 2009.

18. Barrymore Laurence Scherer, "Alice Tully's Pleasing Makeover," *Wall Street Journal*, April 14, 2009, D7.

19. Marco De Michelis, "Modernity and Reform: Heinrich Tessenow and the Institut Dalcroze at Hellerau," *Perspecta* 26 (1990): 143-70.

20. Anthony Tommasini, "At Last Heavenly Acoustics Are Heard in the Hall," *New York Times*, February 23, 2009, C1, 6.

21. Scherer, "Alice Tully's Pleasing Makeover."

22. Alan Rich, "I Love You, Alice B. Tully," *New York Magazine*, September 29, 1969, 50.

23. Harold C. Schonberg, "Philharmonic Hall, and the Listening Is Easy," *New York Times*, October 12, 1969, D19.

24. Allan Kozinn, "Tully Hall Makeover, Warts and All," *New York Times*, July 2, 2009, C1.

25. J. Christopher Jaffe, email to the author, July 31, 2009.

26. Eugene Drucker, conversation with the author, September 11, 2009.

27. Joseph Wechsberg, "A Question of Reverberation," *New Yorker*, November 5, 1955, 100-126.

28. David Taylor, conversation with the author, September 21, 2009, for this and other quotes.

29. Young, interview, 94.

30. *The Philip Johnson Tapes: Interviews by Robert A. M. Stern* (New York: The Monacelli Press, 2008), 161, 151.

31. John Mazzola, interview, January 15, 1992, "Oral History Project," 155.

32. Young, interview, 121.

33. Philip Johnson, interview, August 23, 1990, "Oral History Project," 80.

34. J. Christopher Jaffe, conversation with the author, March 25, 2009.

35. Johnson, interview, 107, 108.

36. Including installing a reflective curve at the back of the stage, breaking up flat wall surfaces in the auditorium, and adding absorptive materials.

37. Harold C. Schonberg, "City Opera Company Sparkles in Its Rich New Setting," *New York Times*, February 23, 1966, 42.

38. See note 36; J. Christopher Jaffe, email to the author, October 11, 2010.

39. Paul Kellogg, conversation with the author, April 3, 2009.

40. Anthony Tommasini, "Mikes Banished, Natural Sound Returns to City Opera," *New York Times*, November 5, 2009, C2.

41. Jane Meyer, "Covert Operations," *New Yorker*, August 30, 2010, 45-55.

42. Raj Patel of Arup, conversation with the author, January 23, 2010, and email to the author, May 30, 2011.

43. Tommasini, "Mikes Banished"; and George R. Steel, conversation with the author, November 25, 2009.

44. Anthony Tommasini, "Improved Acoustics Benefit City Opera," *New York Times*, December 1, 2009, C1.

45. Johnson, interview, 44.

46. *Philip Johnson Tapes*, 164.

47. Johnson, interview, 83.

48. Tommasini, "Mikes Banished."

49. This and the following descriptions of Abramovitz's concept are based on Max Abramovitz, interview, July 18, 1990, "Oral History Project," 140-41.

50. Szell's animosity toward Avery Fisher Hall was attributed to his contretemps with a young acoustician from Bolt, Beranek and Newman; Hans Fantel, "Back to Square One for Avery Fisher Hall," *High Fidelity and Musical America*, October 1976, 70-80.

51. Leo Beranek, *Concert Halls and Opera Houses: Music, Acoustics, and Architecture* (1962; New York: Springer, 2004).

52. Leo Beranek, conversation with the author, October 9, 2009, for this and other quotes unless otherwise stated.

53. Carlos Moseley, interview, May 22, 1991, "Oral History Project," 61.

54. Leo Beranek, interview, October 4, 2001, "Oral History Project," 28.

55. Harold C. Schonberg, "The Occasion," *New York Times*, September 24, 1962, 32.

56. Arthur Gelb, "Acoustical Remedies Set for Philharmonic Hall," *New York Times*, September 25, 1962, 32.

57. Beranek and his group completed the acoustic design of Alice Tully Hall and the Juilliard halls, but credit was given to Keilholz, who supervised construction; J. Christopher Jaffe, whose information came directly from Rudolph Bing (who had interviewed Beranek for the Metropolitan Opera job), email to the author, July 12, 2011, and Carlos Moseley, email to the author, August 15, 2011.

58. Bruce Bliven, Jr., "A Better Sound," *New Yorker*, November 8, 1976, 51-135.

59. Schonberg, "Philharmonic Hall."

60. Amyas Ames, interview, October 17, 1990, "Oral History Project," 70.

61. Avery Fisher, interview, June 12, 1990, "Oral History Project," 35.

62. Carlos Moseley, interview, June 18, 1991, "Oral History Project," 68-70.

63. Fantel, "Back to Square One."

64. Harold C. Schonberg, "New Fisher Hall Opens on an Acoustical High Note," *New York Times*, October 20, 1976, 1, 56.

65. The work was done by Artec and JaffeHolden Scarborough at a cost of $3.5 million.

66. John Deak, conversation with the author, October 31, 2009.

67. Joseph Alessi, conversation with the author, November 13, 2009.

68. Beranek, interview, 46, 48-49.

69. Leo Beranek, *Riding the Waves: A Life in Sound, Science, and Industry* (Cambridge, Mass.: MIT Press, 2008), viii.

70. Anthony Tommasini, "Lincoln Center: Mixed Reviews," *New York Times*, May 10, 2009, 1.

71. Alex Ross, "Waking Up," *New Yorker*, October 19, 2009, 96-97.

72. James R. Oestreich, "Schubert and Mahler, Together Again," *New York Times*, October 23, 2009.

73. Description of the Foster project is based on Zarin Mehta, conversation with the author, May 6, 2009, and J. Christopher Jaffe, conversation with the author, November 24, 2009.

74. Mehta, conversation, May 6, 2009.

75. For the approach of one theater consultant, see Richard Pilbrow, *A Theatre Project* (New York: Plasa Media, 2011).

76. Carl J. Rosenberg, principal of Acentech, email to the author, July 26, 2011.

77. Hugh Hardy, interview, December 9, 1994, "Oral History Project," 36.

Chapter 3
The Present: "If You Can Step on Something You Feel You Own It"

1. Craig Dykers, conversation with the author, September 12, 2008, for this and other quotes.

2. Elizabeth Diller, as quoted in Robin Pogrebin, "At Lincoln Center, Information Is Architecture," *New York Times*, September 2, 2010, C1, 5.

3. "Art Works," National Endowment for the Arts NEA Research Note #102, April 2011.

4. These small parks, like the one behind the Concertgebouw in Amsterdam, have long since been eaten up by the cities surrounding them.

5. See "Electronic Architecture," in J. Christopher Jaffe, *The Acoustics of Performance Halls: Spaces for Music from Carnegie Hall to the Hollywood Bowl* (New York: W.W. Norton & Company, 2010), 171-86.

6. Frank Gehry, introduction to his studio, Yale School of Architecture, April 30, 2010.

7. Tom Remlov, conversation with the author, October 1, 2008.

8. Suzanne Stephens, "Foster and Partners Tethers the Billowing Steel Sage Gateshead Concert Hall to the Banks of the Tyne in England," *Architectural Record*, August 2005, 106.

9. Robert Essert, email to the author, November 8, 2009, for this, other quotes, and the account of Graeme Jenkins's input.

10. Robert Essert, conversation with the author, October 25, 2009.

11. Anthony Tommasini, "Verdi's Moor, Edgy in Cyprus or Dallas," *New York Times*, October 26, 2009, C1, 5.

12. Kevin Moriarty, conversation with the author, July 7, 2010.

13. Rem Koolhaas, conversation with the author, June 9, 2010.

14. George R. Steel, conversation with the author, November 25, 2009.

15. Daniel Libeskind, conversation with the author, February 9, 2011, for this, other quotes, and the architect's opinions about the project.

16. Gerhard Brun, project architect, conversation with the author, June 10, 2010.

17. Michael Dervan, "La Bohème," *Irish Times*, June 18, 2010, 18.

18. Accounts of the cost of Disney Hall's construction have been grossly inaccurate: the $207 million budget established in 1998 by the board for construction costs

alone was in fact met. Stephen D. Rountree, president and chief executive officer, Music Center, Performing Arts Center of Los Angeles County, letter to the editor, *Engineering News Record*, April 1, 2010.

19. Frank Gehry, conversation with the author, April 30, 2010; Gehry also felt stone would glow at night, whereas metal goes dark, as stated in Barbara Isenberg, *Conversations with Frank Gehry* (New York: Alfred A. Knopf, 2009), 239.

20. Leon Botstein, conversation with the author, April 25, 2008.

21. James R. Oestreich, "A Trip to Janacek's World, Complete with Shouting Puppets," *New York Times*, August 19, 2003, E5.

22. Frank Gehry, conversation with the author, July 18, 2008.

23. Frank Gehry, presentation at the Museum of Modern Art, New York, June 10, 2010; conversation with the author, October 21, 2010.

24. Yasuhisa Toyota, conversation with the author, February 26, 2011.

25. Matthew Roitstein, conversation with the author, February 4, 2011.

26. Gehry, presentation at MoMA.

27. Michael Tilson Thomas, conversation with the author, October 13, 2009.

28. See Hans Ibelings, *Space for Music* (Amsterdam: A10 Media/City of Amsterdam, 2007).

29. Huang Ruo, conversation with the author, April 7, 2010.

30. Tod Machover, conversation with the author, April 13, 2010.

31. Renz L. C. J. van Luxemburg, conversation with the author, June 9, 2010; email to the author, May 14, 2011.

32. Stefan Zopp, project architect, conversation with the author, September 30, 2008.

33. Esa-Pekka Salonen, conversation with the author, November 16, 2009.

34. Daniel Barenboim, conversation with the author, August 8, 2010.

35. Alejandro Zaera, "Incorporating: Interview with Jean Nouvel," *Croquis* 13, nos. 65–66, entire issue.

36. Marianne McKenna, conversation with the author, February 12, 2010; email to the author, January 26, 2011.

37. Robert Essert, "Sound Space Design: Crafting Exceptional Acoustics for Opera, Music and Theater" (promotional brochure).

38. Robert Everett-Green, "This New Hall Is Decked with Stunning Sound," *Globe and Mail*, September 28, 2009, R1.

39. Lisa Rochon, "Koerner Hall: Gutsy Vision, Great Vibrations," *Globe and Mail*, September 26, 2009, R9.

40. These projects are only a partial list of those included in Toronto's massive SuperBuild cultural program, which was launched in 2000 and completed with the Koerner.

41. UNStudio project text, as cited in "MUMUTH by UNStudio," *de zeen design magazine*, February 19, 2009, http://www.dezeen.com/2009/02/19/mumuth-by-unstudio/. See also Mario Carpo, *The Alphabet and the Algorithm* (Cambridge, Mass.: MIT Press, 2011), 91–92.

42. Georg Schulz, conversation with the author, May 14, 2009; email to the author, August 18, 2010.

43. Herbert Schuch, conversation with the author, December 7, 2010.

44. Markus Hinterhäuser, email to Thomas Bieber, director of the Waidhofen an der Ybbs Music Festival (who translated the German text into English), December 18, 2010.

45. Carl J. Rosenberg, email to the author, April 1, 2010.

46. Gerard Mortier, *Dramaturgie d'une passion* (Paris: Editions Bourgois, 2009), 37.

47. As quoted in Michael H. Miller, "The Soul of a New Museum," *New York Observer*, August 2, 2010, 42, 43, 52.

48. Walter Benjamin, "The Work of Art in the Age of Mechanical Reproduction," *Illuminations* (New York: Schocken Books, 1968), 234–35.

49. Renée Fleming, email to the author, December 5, 2009.

50. Alicia Desantis, "At the Guggenheim, the Art Walked Beside You, Asking Questions," *New York Times*, March 13, 2010, C1, C7; Erica Orden, "MoMA Attendance Hits Record High," *Wall Street Journal*, June 29, 2010, A27.

Chapter 4
China: "Building Big"

1. Stanislaus Fung and Liu Ke, "Learning for a Strange World," *Archis*, 2006, no. 2, 100.

2. Thomas J. Campanella, *The Concrete Dragon: China's Urban Revolution and What It Means for the World* (New York: Princeton Architectural Press, 2008), 14, 180.

3. Robin Pogrebin, "Lincoln Center to Venture into China as Adviser for a Performing Arts Project," *New York Times*, April 29, 2011, C5.

4. Françoise Ged, "Architecture: Des coopérations fructueuses pour les écoles?" *Etudes Chinoises*, unnumbered (2010), 201–16.

5. "China, France Seal Deal on Aircraft and Nuclear Energy, *China Daily*, http://www.chinadaily.com.cn/china/2010-11/05/content_11504701.htm.

6. Costs throughout this chapter are approximate, since they are not tabulated in China as they would be typically in the West, nor do they include all materials and services that may have been donated for political or business reasons.

7. Ole Scheeren, conversation with the author, December 27, 2009.

8. Steven Holl, conversation with the author, April 8, 2010.

9. Carl J. Rosenberg, principal of Acentech, prepared the acoustical evaluations in this chapter (except as noted) during a trip with the author in December 2009-January 2010.

10. Andrew F. Jones, University of California, Berkeley, lecture at the University of Hong Kong, March 25, 2001, http://www.soh.hku.hk/evt_detail.php?ID=273.

11. "China's New Musical Powerhouses," "Ancient Paths, Modern Voices: Inside Carnegie Hall's Festival Celebrating Chinese Culture" (festival program, New York, 2009).

12. Alex Ross, "Symphony of Millions," *New Yorker*, July 7, 2008, 84-91.

13. Yun Jie Liu, conversation with the author, December 27, 2009.

14. Tao Zhu, "Building Big with No Regret" (paper presented at the conference "Red Legacy in China," Harvard University, April 2-3, 2010), 19-21, 24.

15. Tan Dun, conversation with the author, November 2, 2009.

16. Ji-Qing Wang (Chi Ching Wong), conversation with the author, December 30, 2009.

17. For Hangzhou, June Chuang, Petroff Partnership Architects, conversation with the author, December 29, 2009; for Guangzhou, Shu-Chun Lai, program director, email to the author, September 20, 2011.

18. Salary and ticket prices are in flux; these figures reflect conditions in December 2010.

19. Richard Gaddes, conversation with the author, April 14, 2010.

20. Campanella, *Concrete Dragon*, 14.

21. Wang Zhengming, conversation with the author, December 28, 2009.

22. This description is based loosely on Campanella, *Concrete Dragon*, 141-42.

23. Paul Andreu, *L'Opéra de Pékin: Le roman d'un chantier* (Editions du Chêne, 2007), 58. All quotes from this book are the author's translations from the French.

24. Wang Zhengming, conversation, December 28, 2009.

25. Paul Andreu, conversation with the author, January 5, 2010. Unless otherwise indicated, my descriptions of the architect's intentions come from this conversation.

26. Wang Zhengming, conversation, December 28, 2009.

27. Wang Zhengming, conversation, December 28, 2009.

28. Francesca Zambello, email to the author, April 28, 2010; conversation with the author, June 23, 2010.

29. Ross, "Symphony of Millions," 84-91.

30. Ross, "Symphony of Millions," 84-91

31. Warren Mok, conversation with the author, January 3, 2010.

32. Yun Jie Liu and Chunming Mo, violinist, conversation with the author, December 27, 2009.

33. Campanella, *Concrete Dragon*, 140.

34. Hao Jiang Tian, conversation with the author, January 22, 2010.

35. Campanella, *Concrete Dragon*, 65-71.

36. Ji-Qing Wang, conversation, December 30, 2009.

37. Li Nan, conversation with the author, December 30, 2009.

38. Tao Zhu, assistant professor, Department of Architecture, University of Hong Kong, email to the author, November 22, 2009.

39. Mok, conversation, January 3, 2010.

40. Maryvonne de Saint Pulgent, *Le Syndrome de l'Opéra* (Paris: Robert Laffont, 1991), 244.

41. This and subsequent personal statements by Carlos Ott are based on his conversation with the author, May 29, 2008.

42. Campanella, *Concrete Dragon*, 32-33.

43. "Asymmetrical Auditorium," *Abitare*, April 2011, 192-203.

44. Campanella, *Concrete Dragon*, 44-45.

45. Zaha Hadid and Patrick Schumacher, conversation with the author, May 7, 2008.

46. Ken Smith, "*Turandot*, Steely Grace Beneath the Surface," *Financial Times*, May 12, 2010, 12; Smith, "In China, 'Tosca' Is L'Opéra du jour," *MusicalAmerica.com*, May 17, 2011.

47. Huang Ruo, conversation with the author, March 30, 2011.

48. Hadid is one of several architects, including Ben van Berkel, Wolf Prix, and Toyo Ito (to mention only those discussed in this book), whose topological architecture is continuous, without corners. See Lars Spuybroek in conversation with Ludovica Tramontin, "The Architecture of Continuity," in Spuybroek, *The Architecture of Continuity: Essays and Conversations* (Rotterdam: V2 Publishing, 2008), 208-25.

49. Tao Zhu, email.

50. Weijin Wang, associate professor, Department of Architecture, University of Hong Kong, email to the author, November 22, 2009.

51. Paul Goldberger, "A Dash of the Modern amid Mediocrities," *New York Times*, July 2, 1995.

52. Leo Beranek, *Concert Halls and Opera Houses: Music, Acoustics, and Architecture* (1962; New York: Springer, 2004), 363: "The room is warm enough to feel comforting, spare enough to defer to the primacy of the music."

53. Yasuhisa Toyota, conversation with the author, March 4, 2010.

54. Tan Dun, conversation, November 2, 2009.

55. Jianfei Zhu, *Architecture of Modern China: A Historical Critique* (London: Routledge, 2009), 30, 73.

56. Tao Zhu, email.

57. Jianfei Zhu, *Architecture of Modern China*, 129.

58. Arata Isozaki, conversation with the author, March 31, 2010.

59. Toyota, conversation, March 4, 2010.

60. Toyota, conversation, March 4, 2010.

61. Guobin Yang, "China's School Killings and Rapidly Rising Social Anxieties," *International Herald Tribune*, May 15-16, 2010, 8.

Chapter 5
The Future: Near and Far

1. Peter Murray, *The Saga of Sydney Opera House* (London: Spon Press, 2004), xv, 154-55. See also Ken Woolley, *Reviewing the Performance: The Design of the Sydney Opera House* (Boorowa, NSW, Australia: Watermark Press, 2010), 281-83.

2. Coop Himmelb(l)au, *Selected Projects* (Wolf D. Prix/W. Dreibholz & Partner ZT, n.d.), 12.

3. Wolf Prix, conversation with the author, January 22, 2010.

4. Prix, conversation, January 22, 2010.

5. See chapter 4, note 48.

6. "Toyo Ito for Taichung Metropolitan Opera House," edilportale.com, March 3, 2006.

7. "Le Chantier de la Philharmonie de Paris est en panne," Le Monde.fr, http://www.lemonde.fr/imprimer/article/2010/10/01/1418868.html.

8. Laurent Bayle, conversation with the author, March 21, 2008.

9. Jean Nouvel, conversation with the author, December 20, 2010.

10. DW-World.DE Architecture, April 12, 2010, http://www.dw-world.de/dw/article/0.,5460320,00html.

11. Pierre de Meuron, interview, "Die Elbphilharmonie hat mehrere Lecks," *Frankfurter Allgemeine*, May 28, 2010.

12. Jacques Herzog, conversation with the author, May 21, 2008.

13. Christoph Lieben-Seutter, conversation with the author, July 11, 2008.

14. Llàtzer Moix, "Choque de Trenes en Santiago," *Arquitectura milagrosa: Hazañas de los arquitectos estrella en la España del Guggenheim* (Barcelona: Editorial Anagrama, 2010).

15. Peter Eisenman, conversation with the author, November 3, 2010.

16. Moix, "Choque de Trenes en Santiago."

17. In his essay "Digital Scrambler," Eisenman claims that the overall grid is the result of a new digital code related to natural phenomena such as avalanches, tectonic plates, clouds, and coastlines. Peter Eisenman, *Written into the Void: Selected Writings 1990-2004* (New Haven: Yale University Press, 2007), 147-50.

18. The description of the Performing Arts Center theater interior is based on the author's conversation with Mark Holden, the acoustical consultant, March 4, 2011.

19. *OMA/AMO Theaters, 1987-2009* (OMA, n.d.), 267.

20. This and other quotes and references to Rem Koolhaas are from his conversation with the author, June 9, 2010.

21. Koolhaas, conversation, June 9, 2010; Koolhaas, lecture at Columbia University, Wood Auditorium, February 20, 2009.

22. Renz L. C. J. Van Luxemburg, conversation with the author, June 9, 2010.

23. *OMA/AMO Theaters*, 271.

24. Albert Yaying Xu, "Acoustique de la Philharmonie Luxembourg," http://www.philharmonie.lu/downloads/brochure_ouverture.pdf.

25. For this quote and Toyota's opinions, see "Ten Questions to Yasuhisa Toyota," http://www.elbphilharmonie-erleben.de/en/tenquestions-toyota/.

26. This and subsequent information about the Dalian Grand Theater and Opera House acoustics are based on Eckard Mommertz, conversation with the author, April 6, 2011.

27. Account from Christian de Portzamparc to the author, n.d.

28. Washington Fajardo, national subsecretary of culture, conversation with the author, November 3, 2010.

29. Toshiko Fukuchi, email to the author, April 4, 2011.

30. Lieben-Seutter, conversation, July 11, 2008; Lieben-Seutter, email to the author, April 3, 2011.

31. Holden, conversation, March 4, 2011.

32. Lawrence Biemiller, "Much Ado about Costly Art Centers," *Chronicle of Higher Education*, March 13, 2011, http://chronicle.com/article/In-an-Era-of-Campus-Cutbacks/126691/.

33. Over 360 performing arts centers were constructed in the United States between 1994 and 2008; Barbara Jepson, "Cultural Construction Slowdown," *Wall Street Journal*, September 14, 2011, D7.

34. Mike Boehm, "OC Arts Center Hit by Bond Blues," *Los Angeles Times*, August 9, 2008, E8.

35. Vivien Schweitzer, "Survival Strategies for Orchestras," *New York Times*, May 29, 2011, AR16, 20.

36. Michael M. Kaiser, conversation with the author, November 2, 2010.

37. League of American Orchestras, "Audience Demographic Research Review," December 10, 2009, 4, 11, 18.
38. National Endowment for the Arts, 2008 Survey of Public Participation in the Arts, Research Report #49, November 2009, iii, 1. Reviewing this report in 2011, the NEA found aging audiences to be a factor of an aging population and also age to be less influential on participation levels than education and gender; Mark J. Stern, "Age and Arts Participation," www.arts.gov.
39. Statistics based on information from individual institutions in France and from the Ministry of Culture and Communication, http://www.culture.fr/fr/sections; for national statistics, "Enquête annuelle sur les opéras de la Réunion des Opéras de France," Réunion des Opéras de France, 2009, http://www.rof.fr.
40. Hans Neuhoff, "Konzertpublika: Sozialstruktur, Mentalitäten, Geschmacksprofile," Deutsches Musikinformationszentrum in der kulturstadt Bonn, http://www.miz.org/static_de/themenportale/einfueh rungstexte_pdf/03_KonzerteMusiktheater/neuhoff. pdf, 5.
41. Kaiser, conversation, November 2, 2010.
42. See Laura Zakaras and Julia F. Lowell, "Cultivating Demand for the Arts, Arts Learning, Arts Engagement, and State Arts Policy," Rand Research in the Arts, commissioned by the Wallace Foundation (Santa Monica, Calif.: Rand Corporation, 2008), 1.
43. Daniel J. Wakin, "Beyond Concerts, Carnegie Widens Its Playing Field," *New York Times,* January 20, 2011, C3.
44. Daniel Barenboim, conversation with the author, August 8, 2010.
45. Carsten Siebert, assistant to Daniel Barenboim, emails to the author, August 20, 2010, and July 20, 2011.
46. Daniel J. Wakin, "Los Angeles Orchestra to Lead Youth Effort," *New York Times,* October 5, 2011, C1.
47. See Richard A. Peterson and Gabriel Rossman, "Changing Art Audiences, Capitalizing on Omnivorousness," in *Engaging Art: The Next Great Transformation of America's Cultural Life,* ed. Steven J. Tepper and Bill Ivey (New York: Routledge, 2008), 307-42. For fluctuating numbers of "omnivores," see Stern, "Age and Arts Participation."
48. Richard Taruskin, *The Oxford History of Western Music,* vol. 2, *The Seventeenth and Eighteenth Centuries* (Oxford: Oxford University Press, 2005), 223.
49. Stuart Isacoff, "Perhaps Some Schubert with Your Chardonnay?" *Wall Street Journal,* February 16, 2011, D5.
50. John Canarina, *The New York Philharmonic from Bernstein to Maazel* (Milwaukee: Amadeus Press, 2010), 97.
51. Jennifer Melick, "Just the Ticket," *Symphony,* November/December 2010, 49.

52. http://www.youtube.com/symphony.
53. Daniel J. Wakin, "Orchestras on Big Screens: Chase Scene Needed?" *New York Times,* November 9, 2010, A1, and Daniel J. Wakin and Kevin Flynn, "A Metropolitan Opera High Note, as Donations Hit $182 Million," *New York Times,* October 11, 2011, A1. An important component of the Metropolitan Opera's initiative is the installation of state-of-the-art sound systems in places where the broadcasts are shown.
54. *OMA/AMO Theaters,* 271.
55. David Pountney, conversation with the author, June 29, 2008.
56. Zachary Wolfe, "Arena Opera, Mortier Style," *New York Times,* July 17, 2011, AR1, 16.
57. Valerio Maria Ferrari, http://www.vmcfatelier.com/.
58. The "Space Stage" was shown for three weeks at the International Exhibition of New Theater Techniques in Vienna as an antithesis to the proscenium stage.
59. Kahle Acoustics; J. Christopher Jaffe, email to the author, February 4, 2011.
60. Gerard Mortier, email to the author, April 6, 2011; Nigel Redden, email to the author, March 9, 2011.
61. Salle9620Modulable:9620Vision9620Modulable. webarchive, February 8, 2011.

Photography Credits Numbers refer to page numbers. Peter Aaron/Esto: 128, 129 left; Max Abramovitz sketch courtesy of Thaddeus Crapster, a former designer with Harrison & Abramovitz: 90 right; Sandor Acs; image courtesy of the Archives of Lincoln Center for the Performing Arts, Inc.: 95; Ralph Alberto © FLC/ARS, 2011: 48 left; Paul Andreu: 175, 176, 178, 179, 182, 183, 184; Tom Arban Photography, Inc.: 149 right, 150; Architekturbüro Edgar Wisniewski: 45 bottom; Archives of Lincoln Center for the Performing Arts, Inc.: 78 right; Arte Charpentier Architectes: 187; Atelier Christian de Portzamparc: 56 bottom left, 221 bottom, 222; Atelier Hollein/Sina Baniahmad: 158, 159; Ateliers Jean Nouvel: 146 left, 227, 228, 229; Austrian Museum of Applied Arts/Contemporary Art (MAK): 39 bottom; Avery Architectural and Fine Arts Library, Columbia University: 68, 92; Iwan Baan: 73 right, 112 left, 113, 115, 131, 133, 140 right, 141, 154 right, 196, 197 top, 198, 199; Simone Baldissini-CISA A. Palladio, Vicenza, 2008: 21 right; Richard Beacham, King's Visualization Lab, King's College London: 78 left; Gianni Berengo Gardin: 50 right; Virgile Simon Bertrand: 197 bottom; Bibliothèque nationale de France: 27, 154 left; Bildarchiv Preussischer Kulturbesitz/Art Resource, NY: 37 left; Hélène Binet: 108 left; Martin Blazeby for the King's Visualization Lab, King's College London: 18; O. Bohm-Venice: 10; Eric Boman: 72 left; Nicolas Borel: 56 bottom right; bpk, Berlin/Kupferstichkabinett, Staatliche Museen, Berlin/Volker-H. Schneider/Art Resource, NY: 37 right; BSAE Superintendency of Parma and Piacenza: 22 left; Nicolas Buisson: 108 right; Iñigo Bujedo Aguirre: 236 right; Carlos Ott/PPA Shanghai Office: 193; Craig Chesek/courtesy of Carnegie Hall: 58 right; Coop Himmelb(l)au: 215, 216, 217; Joseph Costa: 90 left; D.R. Chinese newspaper: 181; DeA Picture Library/Art Resource, NY: 9 left; Marc Deville: 186 left; Diller Scofidio + Renfro: 64, 71, 72 right, 73 left, 74 right, 75, 77, 79, 80; Gaston Dubois; image courtesy of the Archives of Lincoln Center for the Performing Arts, Inc.: 85 left; Eisenman Architects: 235, 237, 238; Ennead Architects: 58 left; Esterhazy Provatstiftung, Eisenstadt Palace, Manfred Horvath, Vienna: 32 right; James Ewing Photography: 255 left; Valerio Maria Ferrari: 256; Fisher Dachs Associates, Theatre Planning and Design: 55; FLC/ARS, 2011: 48 bottom right; Fondo Maurizio Sacripanti, Archivio del moderno e del contemporaneo, Accademia Nazionale di San Luca, Roma: 54 center, 54 right; Foster + Partners: 112 right; Reinhard Friedrich (Berlin): 45 top; Gehry Partners: 124, 129 right, 132 left, 134, 249 right; Getty Research Institute, Los Angeles (87‑B14035): 19; Guy Gillette, photographer; image courtesy of Monroe Gallery of Photography and the Archives of Lincoln Center for the Performing Arts, Inc.: 87; Farrell Grehan/Esto: 14–15, 16; Guangdong Xinghai Concert Hall: 203, 204; Steven A. Gunther, courtesy of CalArts: 126 left; Christopher Hagelund/Birdseyepix.com: 107; Hangzhou QianJiang New City Administrative Committee: 164–65, 189, 190, 191, 192; Luke Hayes: 161; Henning Larsen Architects: 246 top left, 246 bottom left; Herzog & de Meuron: 231, 232, 233, 247 top left, 247 bottom left; Ken Howard/Metropolitan Opera: 62–63; Eduard Hueber/archphoto.com: 149 left; Timothy Hursley: 2, 111 top, 116; IBA/Information Based Architecture: 195; Arata Isozaki: 202 right; Katya Kallsen © President and Fellows of Harvard College: 52; Nelson Kon: 221 top; Charlie Koolhaas; courtesy of OMA: 139; Luuk Kramer fotografie: 135, 138 right; Sara Krulwich/The New York Times/Redux: 85 right; KUG/Wenzel: 155 bottom; kuramochi+oguma: 224, 225, 226; Kuwabara Payne McKenna Blumberg Architects: 148; La Scala Theatrical Museum archive: 25 right; Leander Lammertink: 40, 41 top; Erich Lessing/Art Resource, NY: 26; David Levene/Guardian News & Media Ltd: 39 top; Yan Liang: 186 right; Duccio Malagamba: 234, 236 left; Colette Masson/Roger-Viollet/The Image Works: 56 top left; Paul Maurer: 177; Adam Mørk: 137, 138 left; Grant Mudford, courtesy of Los Angeles Philharmonic and Gehry Partners, LLP: 126 right; Musikverein, Vienna: 38; Nationalarchiv der Richard-Wagner-Stiftung, Bayreuth: 28, 29 left; The New York Times/Redux; image courtesy of the Archives of Lincoln Center for the Performing Arts, Inc.: 89; Office of Philip Johnson, Architect; image courtesy of the Archives of Lincoln Center for the Performing Arts, Inc.: 84 right; Stefan Olah; Holzbauer and Partners: 156; OMA: 142, 210–11, 239, 240, 241, 242; OMA/Philippe Ruault: 143; Opéra national de Paris - Opéra Bastille: 59 right; Opéra national de Paris/Christian Leiber: 59 left; Olivier Panier des Touches: 54 left; Richard Payne FAIA: 94; Nicole Petitpierre: 252; Philips, Eindhoven © FLC/ARS, 2011: 48 top right; Rare Books and Manuscripts, Special Collections Library, Pennsylvania State University Libraries: 17, 20, 29 right, 34 right; P.-E. Rastoin; Cité de la musique: 56 top right; Christophe Raynaud de Lage; Photographe Paris: 24 left; Renzo Piano Building Workshop: 50 left, 246 top right, 246 bottom right; Réunion des Musées Nationaux/Art Resource, NY: 23, 24 right; REX: 117; Terry Richardson: 255 right; Christian Richters: 140 left, 153; Roger-Viollet/The Image Works: 25 left; Royal Albert Hall © Chris Christodoulou: 34 left; Royal College of Music, London: 32 left; Philippe Ruault: 9 right, 100–101, 145, 146 right, 147; Henry Salazar: 123; Scala/Art Resource, NY: 21 left; Shanghai Grand Theatre: 185; Noah Sheldon: 127; Shinkenchiku-sha: 200, 201, 202 left; Carsten Siebert: 249 left; Snøhetta: 109; Stadtarchiv Leipzig: RRA (F)1264: 33 left; Stadtgeschichtliches Museum, Leipzig: 33 right; Christian Steiner, photographer; image courtesy of the Archives of Lincoln Center for the Performing Arts, Inc.: 74 left; James Steinkamp, Steinkamp Photography 2011: 31; Steven Holl Architects: 247 top right, 247 bottom right; Stockhausen Foundation for Music, Kürten, Germany: 49; Ezra Stoller/Esto: 65, 83, 84 left; Studio Daniel Libeskind: 121; Studio Daniel Libeskind & Ros Kavanagh: 119, 120; Craig T. Mathew/Mathew Imaging: 125; Richard Termine for The New York Times: 96; 3XN Architects: 136; Toyo Ito & Associates, Architects: 207; Trustees of Sir John Soane's Museum: 22 right; UNStudio: 155 top; Urban-Think Tank/2011: 250; Claudia Uribe: 132 right; Vanni/Art Resource, NY: 35, 36, 44; Peter Wexler and the New York Philharmonic Archives: 93; Nigel Young/Foster + Partners: 11 bottom; Zaha Hadid Architects: 199 right, 218, 219, 220